Exotic Tropicals of Hawaii

Heliconias, Gingers, Anthuriums and Decorative Foliage

Text by **Angela Kay Kepler**
Photography by **Jacob R. Mau**
Photographic Styling by **Brooke Bearg**

Mutual Publishing
Honolulu, Hawaii

Indonesian Wax Ginger

Text ©1989 by Angela Kay Kepler
Photographs ©1989 by Jacob R. Mau except those listed below
which are copyrighted by their respective photographers.

Mutual Publishing
1127 11th Avenue, Mezz. B
Honolulu, Hawaii 96716
Ph: (808) 732-1709 • Fax: (808) 734-4094
Email: mutual@lava.net
Printed in Australia
Fifth Printing January, 1996

All photographs by Jacob R. Mau except

Dave Boynton: 86 bottom, 95 bottom right. **Michael Fogden**: 3, 26 top right. **Kaanapali Beach Hotel**: 46 bottom right.
A. K. Kepler: 8 all except bottom left, 9 all except bottom left, 51 top right, 58 top, 62 bottom right, 63 bottom right, 70 top left, bottom right, 71 bottom right, 85 bottom, 89 top left, 91 top left, right, 92 top, 93 bottom right, 95 top, 103 all except bottom right, 104 top left, right, 106. **Cameron Kepler**: 9 bottom left, 81 bottom left, 85 top right, 107. **P.W. Sykes, Jr**: 8 bottom left.

Long-tailed hermit feeding and pollinating on *Heliconia mathiasii*, Costa Rica.

TABLE OF CONTENTS

Red Christmas Heliconia

ACKNOWLEDGEMENTS

It is always a pleasure to cooperate with people whose living revolves around flowers. This book was produced in collaboration with Alii Gardens, Nahiku, Maui, but special thanks are due to many hobbyists and commercial growers who welcomed us into their gardens and shared their knowledge: Mike Adams, Na Mala Farm; Howard Cooper, Helani Gardens; John Duey, Iao Valley Gardens; Calvin Hayashi, Midori Farms; Tim Lazinski, and Henrietta Chong, Hawaii Protea Co- operative; Mr. and Mrs. Ray Kawabe; Lenore Knobel and Cindy Lawrence, Hawaii Tropical Flowers, Inc.; Hamilton Manley and Rozak Bisel, Sunshine Farms; Eric Tanouye, Green Point Nurseries; Sadamu and Yoshie Okuni; Clayton Uyehara, Dan and Patty Omer, Hana Tropicals.

Birds-of-paradise

We sincerely thank all who assisted with nomenclature, commented on the manuscript, or helped in diverse ways over the years. These include Fred Berry, Ray Baker, John Blumer-Buell, Peter and Claudia Cannon, Richard Criley, Clark Hashimoto, Heliconia Society International, Hotel Hana-Maui, Bennett Hymer, Carl Lindquist, Richard Lyday, John Kress, Exotica Design Maui, Bob Reeds, Sandra Shawhan, Edward Tremper, Tom Wood, and Keith Woolliams. The excellent slide processing was done by Maui Custom Color. We are also most grateful to Michael and Patricia Fogden of Costa Rica for the outstanding photographs on pages 3 and 26. David Boynton, of Kaanapali Beach Hotel, and Paul Sykes also kindly allowed us to use their color slides.

Special thanks are due to the U.S. Fish and Wildlife Service, and to my husband, Cameron, who jointly provided me with the opportunity to live and travel in tropical American rain forests for three years. Cam also provided photographs, encouragement, and manuscript comments. Bonnie Fancher provided secretarial assistance.

A big thank you goes to Brooke Bearg (Brooke Design and Tropicals, Inc.), Jacob's artistic helper for five years; to Howard Cooper, the "father" of Hawaii's heliconia industry, whose support included spending many hours finding perfect blooms for photos; and especially to Alii Chang of Alii Gardens and Tropicals, in Hana, an energetic, generous islander who funded the photography and provided flowers and encouragement. Without him this book would not have been written.

Lastly, I would like to express my personal *aloha* to Jacob Mau for his positive energy, perfectionism and artistic nature, which spawned these exceptional pictures.

INTRODUCTION

With their ravishing elegance, sparkling color, bold geometric lines, and symmetrical perfection, the flowerheads of exotic tropicals furnish the ultimate in floral brilliance.

"Tropicals" is a new word crossing people's lips from Hawaii to Europe. It includes heliconias, gingers, anthuriums, miscellaneous ornamentals such as birds-of-paradise, calatheas, and fanciful "jungle foliage." The word *exotics,* meaning "from elsewhere," also includes common tropical flowers such as plumeria, hibiscus, and bougainvillea (see *Trees of Hawaii,* by Angela Kay Kepler).

Tropicals include an outstanding variety of bizarre flowers: "shish kebabs" of boiled lobster pincers, wiggly strings of gaudy bird's beaks, floral necklaces of pink and white porcelain "shells," waxy golden "beehives," translucent sculptures seemingly molded from glacial ice, a "rattlesnake tail,"

Torch gingers rest within a feathery papyrus bed.

and even bunches of tiny pink bananas that automatically peel back their skins.

Since their widespread availability in the early 1980s, these flowers, which are one of the fastest-growing segments of Hawaii's diversified agriculture, are capturing larger audiences each year. They are valued for their stunning shapes and colors, long vase life, immaculate quality, and versatility. Most abundant from September to May, they appeal primarily to sophisticated corporate, professional, and private buyers between 30 and 45 years old. In New York, 1987 sales of tropicals accounted for approximately 35 percent of the floral market, despite the fact that a single bloom can cost up to $50. In Hawaii, business in tropical exotics is mushrooming; in 1987, sales approached $16 million.

Ranging from six inches to six feet long in size, the compelling colors and bold lines of tropicals grace hotel lobbies, restaurants, conference rooms, offices, and private homes. Large ones are unsurpassed for massive displays. Trend-setting interior design magazines and contemporary furniture showrooms often use tropicals instead of traditional flowers. Though prices vary considerably, they are relatively expensive outside Hawaii. Reasons for this include the competition between productive growing space and Hawaii real estate prices, among the highest in the world. Some tropicals take two years to flower, and most require a rain forest environment. Large plants need special attention as rains and winds knock them over easily. Extra labor is necessary to free the blooms of debris and to disinfect them before shipment by air freight, which isn't cheap.

Although the Hawaiian Islands are themselves showcases of evolution, having 960 native species, 91 percent of which are not found anywhere else, Hawaii also claims top place in the world for the largest number of introduced plants. Most island landscaping uses ornamentals selected from over 5,000 imported plant species. Small quantities of heliconias and gingers have beautified Hawaiian gardens for decades, and some now grow wild along

Hawaii's lush windward roadsides. Since the 1970s, horticultural expertise from Hawaii's multiracial population — Asian, European, Pacific and North American mainland — has engendered a dramatic interest in tropical floriculture. This has resulted not only in a booming floral trade, but in an increasing variety of ornamentals planted in resorts and botanical gardens, and the preponderance of gorgeous flower arrangements in public places. The major growing areas are Hana, Maui; Hilo to Mountain View, on the Big Island; and a few valleys on Oahu, where growing conditions simulate tropical rain forests: consistent warm temperatures, muted light intensity, and rainfall exceeding 75 inches per year.

This book covers 136 species of hybrids and cultivars of tropicals that you are most likely to encounter in Hawaiian gift boxes and floral displays, and at florists and flower farms. It also covers most flowers and foliage grown in, and marketed from, Florida, the Caribbean, Central America, northern Australia and Samoa. Because of past and current confusion with nomenclature, I have attempted to consolidate as many names as possible: marketing, horticultural, trade, nursery lingo, common English, and scientific names. Thus, a special feature of this book is a compilation of correct, up-to-date common and scientific names for each species, formulated in collaboration with the Heliconia Society International, American Ginger Society, and the Lyon Arboretum, on Oahu.

In whatever manner you encounter these exquisite plants — offspring from the earth's vast equatorial forests — I hope they will deliver warm feelings to you from Hawaii's very special islands. May they also deepen your understanding of Life — its harmony, its varied means of expressive beauty, its constant change, and above all, the miracle of its underlying energy.

Kay Kepler

Pink ginger (*Alpinia purpurata* cv. 'Eileen McDonald'), an import from Tahiti, is now popular in landscaping.

Natural Habitats of Tropicals

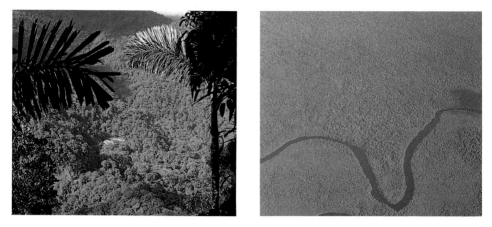

Left: Rancho Grande National Park, Venezuela, like all rain forests, is composed of multistoried layers stretching upwards from carpets of ferns and seedlings to rounded treetops 100-150 feet high. The grandeur of these arboreus cathedrals, bristling with bromeliads, orchids and vines, dwarfs even a "giant" heliconia, 20 inches high. *Right:* One hectare (2.2 acres) of virgin forest in Guyana, such as along the Mazaruni River, contains 178 species of trees; in Maryland you'd be lucky to find 15 in an equivalent area.

Left: Bromeliads are abundant in tropical America. From deserts to drippy cloud forests, they cluster in rosetted, aerial gardens on branches and tree trunks. Here "Flaming Torch' (*Guzmania berteroniana*) embellishes Sierra palms in the Caribbean National Forest, Puerto Rico. *Right:* In glorious full bloom, advertising nectar to an abundance of iridescent hummingbirds, thousands of Vriesea Bromeliads grow in an elfin forest that surrounds the continuously active volcano, Poas, in Costa Rica.

Left: The natural haunts of tropicals — rain forests, considered to be the evolutionary cradles of life — are staggeringly complex and ineffably beautiful. Malaysia, rich in gingers and depicted in a misty sunset at Fraser's Hill, is home to a mind-boggling 20,000 species of plants. Tropical America, the center of evolution for heliconias, bromeliads, and tropical foliage, harbors similar botanical diversity. ***Right:*** La Virgen del Socorro, Costa Rica, is as yet unprotected. Not far away lies La Amistad International Park and Biosphere Preserve (418,000 acres), a product of friendship between Costa Rica and Panama. Though still largely unexplored, its faunal diversity in 1989 included 250 species of amphibians and reptiles, 215 species of mammals, 115 species of fish, approximately 10,000 species of insects, and 560 species of birds. For comparison with the last figure, the entire United States and Canada host around 750 bird species.

Left: The author pauses beside a wild rattle-shaker (see page 77), Penas Blancas Valley, Costa Rica. ***Right:*** Heliconias, Braulio Carrillo National Park, Costa Rica.

A dazzling bouquet of large and small heliconias is rounded out by folded *ti* leaves.

HELICONIAS

Heliconias — glorious splashes of floral color that dot humid rain forests and gardens of tropical America, Asia and the western Pacific — are now grown in quantity in Hawaii's warm, muggy lowlands. Their strikingly elegant flowerheads arise within banana-like clumps of oval leaves that range from 2 to 20 feet high. It is important to note that their leaves customarily rise directly from ground level rather than from an erect trunk; you can waste a lot of time looking for heliconias if you cannot distinguish them from bananas. Formerly classified with bananas, heliconias are now in a separate family, Heliconaceae. They never produce banana-like fruit.

The name *Heliconia* (pronounced "hell-ee-cone-ee-uh") honors Mt. Helicon, home of the ancient Greek gods. Depending on which botanist you follow, there are between 250 and 400 species, 98 percent of which are native to tropical America. Their leaves were used in food storage, cooking and house construction, and obviously the blooms were used for decoration. In the following pages, I have attempted to use everyday language. However, as heliconias are constructed peculiarly, it is necessary to clarify an unfamiliar word: *bract*.

The dramatic color and form of a heliconia bloom is not due to its flowers, but rather to highly modified leaves called bracts. Think of bracts as "floral boats," curving gracefully upwards in alternating patterns along a thick stem. Each "floral boat" houses the heliconia's true flowers — rather inconspicuous, curved tubes of white, yellow, pink, orange, or green hue. The flowers may curve up or down. Hence we cannot call the heliconia bloom a flower; it is a flowerhead — a cluster of many bracts, each containing flowers. The same principle pertains to artichokes, proteas, and banksias.

Many heliconias available in Hawaii are identical to their wild counterparts; natural hybrids are rare, and horticultural hybridization, except for psittacorums, is relatively new. Visit any tropical American rain forest and you will encounter heliconias everywhere within their majestic, green depths, and also brightening clearings, roadsides, and streambanks. People also plant the

Forrest Koa Chang with a cream variety of yellow caribaea (*Heliconia caribaea* cv. 'Cream').

common species in gardens. Have binoculars handy — right at dawn, when heliconias produce the most nectar, you may spy a hummingbird shimmering in a streak of sunlight, unintentionally pollinating the plants as it zips from flower to flower, sipping nectar through its tiny, straw-like tongue. Heliconias diminish nectar production after mid-morning so that ants and bees won't steal it from the birds!

In this chapter, heliconias are organized into large, medium, and small erect flowerheads, and pendent flowerheads. Characters that aid in identifying them are bract shape, degree of bract overlap, the angle between bract and stem, and, of course, whether the flowerhead is upright or pendent.

YELLOW CARIBAEA (*H. caribaea* cv. 'Cream,' left) and RED CARIBAEA (*H. caribaea* cv. 'Purpurea,' right). With care, their neatly symmetrical flowerheads may last three weeks.

Erect Heliconias

Yellow and Red Caribaeas

Heliconia caribaea

Other Names: *giant yellow heliconia, gold caribaea, Caribbean heliconia, caribbea, yellow carib; pale yellow seedling: caribaea 'Cream'; large red: 'Purpurea'; small red: 'Richmond Red,' 'Purpurea 2' (hydrid of* H. caribaea *and* H. bihai).

Glowing with sunshiny waxiness, this magnificent heliconia is one of the largest. Each golden-yellow or scarlet flowerhead measures up to eight inches wide and 16 inches long, while its leaves may attain a length of 20 feet. Originally from the Caribbean, it was intro-

duced to Hawaii in 1958. In its native islands (Puerto Rico, Jamaica), huge clumps of its banana-like plants beautify streambeds, plunge pools below waterfalls, roadsides, and rain forest clearings. Its bracts, each about five inches long and tipped with green, are arranged in two dimensions. Their broad, curvaceous bases overlap like shingles, producing a very compact, symmetrical appearance. Caribaeas, despite their size, enjoy extended blooming periods: fall, winter and spring, a fascinating evolutionary development. In their natural haunts, each bloom lasts several weeks and acts as a long-lived visual signal to attract birds whose territories are large.

Florida horticulturalists are experimenting with this splendid banana-like heliconia in its various color forms and hybrids—red, yellow, chartreuse, greenish-

A close-up of 'KAWAUCHI' reveals charming green-and-white flowers.

yellow, gold, scarlet, and maroon—to produce an indoor plant. It may not grace a tiny living room, but what a regal greeting to a hotel or a dentist's waiting room.

Left: 'FLASH', formerly thought to be a hybrid of red and cream forms (*H. caribaea* cv. 'Flash') is patterned variably, yet breeds true. ***Right:*** Japanese-named 'KAWAUCHI', (*H. caribaea* cv. 'Kawauchi') is less massive than its close kin and is characterized by narrow, waxy bracts with golden margins.

RED BOURGAEANA, one of the few heliconias with a sateen sheen.

Red Bourgaeana

Heliconia bourgaeana

A modern arrangement features 'Splash' and other tropicals featured in this book. *Courtesy of Alii Gardens.*

Bold and satiny, red bourgaeana's (pron. "bore-zhay-ah-nah") huge flowerheads today adorn offices, hotels, and conference centers not only in Hawaii, but in North America and in Europe and Asia. Available in spring, it is perfect for large-scale flower arrangements that last up to a month.

In Hawaii, many red heliconias are grown commercially. Some are difficult to differentiate, but this one is easy. Its color, not truly red, is a shiny rose-red with purplish sheen. Red bourgaeana's large flowerheads, measuring one to two inches long, have a thick yet translucent texture. Their bracts are fat and rounded, with bases barely overlapping. Note their absence of green margins.

Red bourgaeana's huge leaves (up to 14 feet high), like those of all heliconias and bananas, are designed to capture as much light as possible. The amount of energy required to produce huge fleshy flowerheads and copious fruits is prodigious.

'SPLASH' (*H. champneiana* cv. 'Splash,' left) and 'HONDURAS' (*H. champneiana* cv. 'Honduras,' right) are very closely related. Has an artist been spattering paint around with a toothbrush, or, as the contemporary Hawaiian name, *huamoa,* suggests, did some "blood from the chicken yolk" find its way onto 'Splash's' shiny bracts?

Rainbow Heliconia

Heliconia wagneriana

Other Names: *pink-spotted lobster claw, wagneriana, Easter*

The bright greenish-yellow bracts of rainbow heliconia, daubed with sizable rouge "cheeks," are a visual feast. Its long-pointed bracts arch with a particularly pleasing curvature. Its fan-shaped clumps, up to 12 feet high, thrive as well in Hawaii as in its native Costa Rica and Panama. Full sun, ample water, warm air, high humidity, and good soil are all it needs. However, its blooming season, March to May, is short; current research is aimed at expanding its period of availability. Rainbow heliconia resembles giant lobster claw (see page 19). However, its unopened bracts form a jagged, herringbone pattern, its "cheeks" are smaller, and its yellow coloration is greenish rather than bright.

Large Lobster Claws

*Heliconia bihai**

    Appropriately named, these striking exotic flowerheads mimic shish kebabs of boiled lobster pincers. In true crustacean fashion, these crimson claws are rigid, retaining their

COMMON LOBSTER CLAW (lobster claw red, humilis), originally from Brazil, Grenada and Trinidad, is well-known in Hawaii. Besides its use in landscaping and in the floral trade, lobster claws are also naturalized in some windward areas, for example, along Maui's Hana Highway. Locals cut its flower stalks and oval leaves to decorate for special events.

* Many books use the obsolete name H. humilis, *now incorrect except as an old name for* H. psittacorum.

shape and color for weeks. Just as lobsters from different seas come in an array of shapes, sizes, and colors, so also do their floral namesakes.

The most diverse of all heliconias, lobster claws are highly variable and intergraded, perhaps because their many Caribbean island forms never became sufficiently isolated from each other genetically. Red, gold, orange, maroon, and green, singly or in combination, comprise this protean assemblage. Their bright bracts, edged in green, average five to seven inches wide. Entire flowerheads, without stems, attain 16 inches in length, some of the longest of all heliconias. The national flower of Trinidad is a lobster claw.

The first heliconia to be chronicled — the red form of *Heliconia*

Above: Beverly Akiona sorts a fresh harvest, including an armful of COMMON LOBSTER CLAWS.

Right: Almost all the world's heliconias evolved colored bracts, but a few, such as this 'EMERALD FOREST' LOBSTER CLAW (*H. bihai* cv. 'Emerald Forest') from St. Lucia in the Caribbean, emerged green. Its bracts do not have long, curled-under tips, a detail which easily distinguishes it from the Asian/Pacific laufau (page 26).

Below: A 'YELLOW DANCER' (*H. bihai* cv. 'Yellow Dancer') originated on St. Vincent. Small forest birds, the major pollinators of heliconias in natural and adoptive habitats, occasionally visit the open bracts for water as well as nectar.

18

For years thought to be a separate species, the patterned blend of red, yellow, and green in 'GIANT LOBSTER CLAW'* (*H. bihai* cv. 'Giant Lobster Claw') is yet another variant of the common lobster claw. Imported from Brazil to Maui in the early 1970s, this large exotic is now a much-esteemed cut flower. Don't confuse it with rainbow heliconia (p. 16).

* *This species is incorrectly called "choconiana," a confusing name also applied to* H. aurantiaca *(p. 31) and* H. psittacorum *(p. 32).*

bihai — was described by a Monsieur Plumier in 1703. Understandably, he classified it with the closely-allied bananas. The Spanish name, *platanillo* (little wild banana) also reflects this true genetic affinity.

In Hawaii, lobster claws sprout a prolific array of flowerheads most of the year, peaking in summer. For a few weeks in mid-winter the old ragged foliage dies back.

Hawaii's ancient royalty, the *alii,* are internationally known for their fabulously soft feather capes, whose red-and-yellow feathers were painstakingly hand-stitched onto fine fiber nets. Today, red and yellow still symbolize royalty, particularly the Kamehamehas, a succession of influential kings who ruled the islands during the nineteenth century. How fitting, therefore, for a heliconia bearing regal colors to commemorate this touch of island history.

'KAMEHAMEHA' LOBSTER CLAW or stripe (*Heliconia bihai* cv. 'Kamehameha').

You'll find an image of Kamehameha the Great, cloaked in his fine red-and-yellow cape, on the Hawaii Visitors' Bureau signposts at notable scenic spots. His floral allonyms perk up botanical gardens, hotel lobbies, and, perhaps, your own home, if you were the lucky recipient of some beautiful tropicals from Hawaii.

'KAMEHAMEHA' (above right) and 'JACQUINII' (right) are so similar that many people call them both "Kamehameha." Natives of tropical America, both are red-and-yellow striped varieties of the common lobster claw. Their flowerheads may reach 18 inches in length, bearing about ten bracts per stalk. Greenish-white striped flowers arc gently from boat-shaped bracts about five inches long. Note, however, the subtle individualities of bract color: 'Kamehameha' is yellow, sharply divided from the red; 'Jacquinii' is golden, grading less distinctly into the red.

'JACQUINII' LOBSTER CLAW (*Heliconia bihai* cv. 'Jacquinii').

Above: 'Jacquinii' lobster claws ready for cutting, Alii Gardens, Maui. All tropicals, including heliconias such as this, last two to three times longer than traditional flowers.

Left: 'CLAW 2'* (*H. bihai* cv. 'Claw 2'), often confused with the common lobster claw, is trimmer and less wild, with wider spacing between its bracts. Its bright red flowerheads, edged in dark green, average 12 inches long. Spring and summer are its optimal blooming months in Hawaii. Don't confuse its orange form with 'Sharonii' (p. 22).

Smaller Lobster Claws

Heliconia stricta

Slimmer, shorter-stemmed, and more refined in shape than the larger lobster claws (see page 17) are several smaller lobsters whose bracts tend to be flat-sided rather than rounded. These glamorous tropicals are ideal for small arrangements, as their flowerheads range from five to 12 inches long and are not too heavy. Hawaii has no snakes, but in tropical America, the original home of lobster claws, it is not uncommon to find snakes, possums, bats, tree frogs, and spiders making snug homes in heliconia leaves and flowerheads.

Above: Colored a rich dark salmon, edged with yellow and green, 'SHARONII' or dusty rose (*H. stricta* cv. 'Sharonii') is available during summer and winter. Its broad foliage, deep red below, is borne proudly on stiff red stalks. On account of the heavy rainfall in heliconia habitats, the boat-shaped bracts of most species contain mushy "soups," ideal nurseries for mosquito wrigglers, insect larvae and tree frogs. Such semi-permanent mini-ponds often house well-balanced communities of plants and animals. Naturally, this organic matter is removed before local use and shipment overseas.

Left: 'FIREBIRD' (*H. stricta* cv. 'Firebird,' left) and 'TAGAMI' or 'Royal' (*H. stricta* cv. 'Royal,' right) are popular both as cut flowers and garden ornamentals. Notice the yellow central axis in 'Tagami.' Their glossy foliage attains only five feet in height and blooms in spring and summer. Instead of forming leafy clumps like most heliconias, their rhizomes (underground stems) creep like bamboo, prompting the general name "red running heliconias."

The delightful, tiny 'DWARF JAMAICAN' HELICONIA, (*H. stricta* cv. 'Dwarf Jamaican'), only two feet high, makes a perfect ground cover or garden border. It is known variously as dwarf lobster, Jamaica, mini-Jamaica, mini-red 'Jamaica,' and dwarf humilis. Its engaging, five-inch dainty flowerheads are rose-colored, evenly graded from pale to deeper hues. Each bract is ridged with green on its upper edge, matching the tiny green-and-white striped flowers which peek out from their protective "boats."

'Dwarf Jamaica,' introduced into Hawaii in 1971, is easily grown in pots outdoors; hopefully it will soon be available for indoors. A year-round bloomer, its yields peak during winter. Even when not flowering, it is valued for its attractive foliage, which resembles an aggregation of miniature banana plants. Its natural range in Jamaica and Trinidad spans from sea level to 4,500 feet. Thus, in cultivation this tiny heliconia adapts well to variable temperatures. Best flowering occurs, however, when days are sunny and nights cool.

Green Heliconias

Heliconia indica, H. laufau

Other Names: indica, green lobster claw, illustris, ornamental banana plant, paka *(a Fijian species). See* H. bihai *cv. 'Emerald Forest' (p. 18) and* H. caribaea *cv. 'Chartreuse' (p. 12)*

Green heliconias may not present quite the visual appeal as their gaudy cousins, but their lack of color is amply compensated by their ornamental foliage, glossy red fruits, native usage, and fascinating biology. Culturally speaking, the six species of green-flowering Asian and Pacific heliconias, ranging from Samoa to central Indonesia, are utilized more by native peoples than the entire 250 species in tropical America.

First collected around 1750 by Dutch explorers in the jungly Moluccas, which are islands near New Guinea, the GREEN HELICONIA (*H. indica* var. *micholitzii,* left), occurring in several varieties, is the most widespread heliconia in the Old World tropics. For centuries, villagers of many cultures have used it for thatching and for wrapping foods prior to steaming in underground ovens.

Another important use, still in effect, involves shredding the leaf's fibrous midribs into long strands. These fibers are then twisted and sold in markets for squeezing coconut meat into rich coconut cream, a daily staple throughout the coconut's range. In times of famine, villagers of insular nations such as Fiji and Vanuatu (formerly New Hebrides) boiled heliconia flowers and fruits as "starvation food."

The green heliconia grows from six to 18 feet tall and produces large flowerheads 10 inches long and 12 inches wide. Its subdued coloration, best appreciated in combination with colorful heliconias, is due to some fascinating biological phenomena. Asian heliconia flowers are pollinated by bats, which are color-blind; in tropical America they are pollinated by hummingbirds, whose color vision compares favorably with our own. Obviously there is no pressure for plants to produce bright flowers if their pollinators only register black-and-white. Since hummingbirds do not occur in Asia, nectar-sipping bats, their mammalian equivalent, evolved long tubular tongues and a predilection for sweet nectar. Which came first, the heliconia or its pollinators? These plants provide a striking example of co-evolution.

The resplendent leaves of PINKSTRIPE-LEAF HELICONIA (*H. indica* var. *rubricarpa* cv. 'Rubricaulis') are stunning. Growing rapidly to about 12 feet high, each unfurls to expose a huge oval of fine, feather-like striations in rose-pink and green, centering in a bright rose midrib and stalk. Its original scientific appellation and alternate common name, roseo-striata, echo this unusual patterning. Most of the plant's energy is utilized in the production of these fancy leaves; hence, the flowerhead's beauty suffers.

Right: Heliconias also may form temporary homes for bats. White TENT BATS (*Ectophylla alba*) snuggle together under a heliconia leaf in a tropical American rain forest. The bats have chewed both sides of the midrib, causing the leaf to collapse like a tent.

Below: The vivid PINK-VEINED BANANA PLANT (*H. indica* var. *rubricarpa* cv. 'Spectabilis') sports leaves of rose, green, and bronze, depending on age. (The green leaf is oldest.) Notice the close parallel arrangement of the leaf veins, a characteristic feature of the heliconia/banana family. The pink-veined banana plant is native to New Guinea and neighboring island groups. It is also called rubricarpa.

Above, left and right: LAUFAU *(Heliconia laufau)* from Samoa, bears a fan-shaped flowerhead, typical of all Asian and Pacific heliconias. Though pretty in its own right with gold-edged green bracts, laufau's primary culinary use is as a food wrapper. Fish or vegetables are wrapped in its leathery leaves, placed in a hot underground oven, covered with wet leaves and old woven mats, then left for hours. This traditional Polynesian method of cooking, still practiced daily on many islands, parallels the use of banana leaves at Hawaiian *luau* (feasts). Each item of food resembles a Hawaiian *laulau* (made with ti-leaf wrapper) or Mexican tamale (corn-husk wrapper). Note the artistic, whorled arrangement of bracts, this time bearing seeds whose mature orange color will attract birds, such as thrushes, that devour and, consequently, disperse them. Laufau is now available in light and dark green varieties.

Yellow Fan

Heliconia lingulata

Other Names: yellow variety: *lingulata, linqualata;* two red-and-yellow-varieties: *red fan, redtip fan, 'Pagoda,' "birdiana," linqualata birdiana, Birdeyana*

During the 1970s, Howard Cooper — an enthusiastic nurseryman, owner of Helani Gardens, in Hana, Maui, and father of Hawaii's heliconia trade — was attending a conference in Honolulu. Decorating banquet tables was a gorgeous flower arrangement featuring a yellow heliconia with long, narrow bracts arrayed in three dimensions. What was it? Detective work followed, and soon a few precious yellow fans (*H. lingulata* cv. 'Fan') were thriving in Hana. Notice the slender, well-spaced bracts and rather top-heavy look of yellow fan and red fan (both, below left). Both are summer bloomers; 95 percent of all heliconias are seasonal, each species having a peak blooming period. Despite their tropical origin, some are common in winter, such as holiday heliconias (see page 30), and caribaeas (see page 13).

Above: RED FAN (*H. lingulata* cv. 'Pagoda'), marketed on Maui as "birdiana," sprouted spontaneously in an unlikely corner of Helani Gardens. Its bracts, rather than being plain yellow, developed generous washes of orange. Was it a yellow fan pollinated by an insect or bird which had previously visited a red-flowering heliconia? No matter - this handsome tropical is here to stay, the original plant still proudly tended by Howard Cooper.

Latispathas

Other Names: orange variety: *orange latispatha;* yellow variety: *yellow latispatha;* red variety: *'Distans,' red latispatha*

Simple to identify, the arty latispathas come in three varieties: orange, yellow-and-red, and red. Long and tapered, their well-separated bracts radiate, not in two dimensions, but in three. Note that the lowest bract (10 to 12 inches long) sports a small leaf-blade at its tip (*latispatha* means side-blade), disclosing the basic biological fact that even though bracts may look like petals, they are greatly modified leaves.

This tall, rambling heliconia is definitely not for a small backyard. It grows rampantly, attested by its conspicuousness, over a wide elevation range in its native Central America and in northern South America. In a few places in lowland Hawaii, the yellow variety is naturalized in the wild, such as at Puaa Kaa State Park on Maui. Prolific growth and bracts

bursting with juicy fruits bear testimony to its success.

In their natural habitats, plants and birds frequently change to suit each other's needs. Tropical America is a storehouse for this phenomenon of co-evolution. Most heliconias, for example, evolved curved flowers so that curve-billed hummingbirds could pollinate them. Latispathas chose a slightly different evolutionary path. In order to avoid competition with other plants, they developed short, straight flowers in response to the presence of short-billed hummingbirds . . . or were the short-billed hummingbirds able to survive because straight-billed heliconia flowers existed? In Costa Rica alone, 48 species of hummingbirds have provided plenty of latitude for diversification not only in heliconias and bromeliads, but in hundreds of other flowering vines, shrubs and trees as well.

A beautiful cultivar, RED LATISPATHA (*H. latispatha* cv. 'Distans'), bears long golden bracts emblazoned with red. There is also a pure red form. A summer bloomer, it is more appealing and longer lasting than the plain varieties. (Compare all three in the excellent heliconia collection at Waimea Arboretum, Oahu.) Latispathas are superficially similar to yellow and red fans (p. 27), but note the narrow apex of the former as compared to the bulging apex of the latter.

Holiday Heliconias

yellow: *Heliconia angusta* cv. 'Yellow Xmas'
orange: *H. angusta* cv. 'Orange Xmas'
red: *H. angusta* cv. 'Holiday'

Other Names: yellow variety: *yellow Xmas, yellow flava, flava;* orange variety: *orange Xmas, pagoda;* small red variety: *red Xmas, "H. Xmas";* large red varieties: *Xmas grandiflora, large Xmas, "March Xmas"*

The pure, unblemished flowerheads of these unusual heliconias span the period Halloween to Easter, peaking in December and including the two "red" holidays, Christmas and Valentine's Day. These Brazilian novelties add tropical flavor to holiday decorations traditionally dominated by poinsettias and chrysanthemums. Now grown in Hawaii, they symbolize relaxation, fun and romance, especially to inhabitants of wintry climes. Note the striking red-and-white flowers of the red variety.

Researchers have high hopes for the future of these small heliconias, for they are not dependent on high light intensity, they can grow in eight-inch pots, they will tolerate temperatures down to 40 degrees F, and they rarely exceed three feet in height.

Dwarf Golden Heliconia

Heliconia aurantiaca

Other Names: *coat-of-arms**

From Costa Rica and Mexico comes this cheerful relative of parrot's beak and holiday heliconias. Its three-dimensional, five-inch-long orange flowerheads practically explode with yellow, banana-like florets. In its natural habitat the dwarf golden heliconia grows in light gaps within tall forests and along streams. In cultivation, therefore, it is ideal for mass plantings (three to five feet high) beneath treeferns, bananas, taller heliconias or other tropical plants. Note its unusual, "lacquered," stemless leaves. It is most available in spring.

* *On Oahu, coat-of-arms is applied to* H. hirsuta.

Parrot's Beak Heliconias

Heliconia psittacorum

Other Names: psittacorums, false birds-of-paradise

Small, dainty, and distinctively exotic, parrot's beak heliconias make delightful ground covers in warm areas, blooming abundantly all year, in contrast to 95 percent of other heliconias, which are seasonal. They rarely exceed four feet in height. Flowerheads appear painted and glowing with brilliant colors.

In parrot's beak heliconias, the floral bracts arise from roughly the same point on the stem, thus resembling birds-of-paradise rather than conventional heliconias. This superficial similarity has given rise to their oft-used name: false birds-of-paradise. Bright, black-tipped flowers, curving gracefully from within the bracts, add to their overall visual appeal.

The sparkling, multicolored Sassy (*H. psittacorum* cv. 'Kaleidoscope').

The most versatile of heliconias for the homeowner, parrot's beaks do not mind coolish temperatures, and therefore they fare well outdoors in Florida and on the lower mountain slopes of Hawaii, where plots of ten square feet can average 150 to 200 blooms annually. In hot, humid areas, and with a rich nitrogen fertilizer, harvests are greater. As any gardener knows, these figures far exceed yields from true birds-of-paradise. Each cut bloom lasts two to three weeks. Recent experiments have shown that cutting stems before 8 a.m., then recutting every few days, extends their life another week.

Parrot's beak heliconias, like lobster claws, intergrade easily, and thus are the subject of much horticultural attention. Imports and new hybrids are continuously contributing to commercial enterprises. Because they are small and economical to ship, they are fast becoming darlings of the floriculture trade.

Above: Two varieties of parrot's beak heliconias are pictured. The left flowerheads, known as 'PARA-KEET' (*H. psittacorum* cv. 'Parakeet'), show cheery rosy-yellow bracts and yellow flowers. It originally hails from Venezuela and is sometimes called *rhizomatosa*. On the right is PARROT, a named hybrid (*H. psittacorum* x *spathocircinata* cv. 'Golden Torch'). Larger and sturdier than other psittacorums, it is called 'Golden Torch' in Florida, where it is marketed as both an indoor and outdoor landscaping plant. Its rigid "arms," like opaque, golden sun rays, are not a result of evolutionary chance. They have been selectively bred for longevity, color, and durable texture.

Left: 'NICKERIENSIS', also a named hybrid from South America (*H. psittacorum* x *H. marginata*) similar to the parrot, glows even brighter with reddish-orange, rather golden bracts.
Below: Who knows in what city these PARROTS will bring joy to their recipients - New York, Paris, Tokyo?

Above: PARROT'S BEAK (*H. psittacorum* cv. 'Rubra'), introduced to Hawaii from Puerto Rico in 1950, is the most commonly encountered variety in island gardens, featuring reddish-orange bracts and orange flowers. Since it tolerates far less humidity than most heliconias, it thrives in drier areas, and in both sun and shade. It is this variety that conferred the general name, "parrot's beak heliconias," to the numerous cultivars of *Heliconia psittacorum*.

Right: Bursting with rich reds and oranges, 'AN-DROMEDA' (*H. psittacorum* cv. 'Andromeda'), of unknown origin, came to Hawaii in 1978 from the beautiful Andromeda Gardens on Barbados Island in the Caribbean. Although not widespread yet, this vibrant beauty has a high commercial potential. Other new gay cultivars include Lady Di, St. Vincent Red, and Marisol, which claims the distinction of being the tiniest heliconia to date: a mere one and a half inches and capable of growing in six-inch pots. For a heliconia, this is a near miracle.

Above: Mike Adams, a Hawaiian owner of Na Mala Farm, Maui, stands amidst lush 'SASSY' HELICO-NIAS. If you live in a suitable area, treated rhizomes, ready to plant, are available from many farms (p. 96) and will grow well where minimum temperatures do not drop below 60 degrees F. They need not be planted immediately; kept moist and shaded, they last for weeks.

Right: SASSY (*H. psittacorum* cv. 'Kaleidoscope'), a captivating, many-splendored variety, was a mystery plant in Hawaii until the early 1980s. Hand-carried to Maui, it is now a much-loved cut flower and landscaping gem. Vivid orange flowers arise from its apple-green, yellow and pink-hued bracts. In Florida it is known by its cultivar name, 'Kaleidoscope.'

Pendent Heliconias

Some of the most dazzling heliconias have dangling, rather than upright, flowerheads. Obviously the bushes are tall; the following seven species range from six to 15 feet high. Their blooms bring drama and stylish character to lobby, restaurant, or living room decor.

Red Collinsiana
Heliconia collinsiana

Other Names: *Collin's heliconia, hanging heliconia, collinsiana lobsterclaw, red 'Collinsia'*

Pendent, curved triangles of reddish-pink distinguish this graceful heliconia. Its 12-foot-high clumps are composed of spindly, banana-like "pseudotrunks" from which eye-level flowerheads dangle like strings of delicate fruits. Red collinsiana's bracts, six to 12 inches long, are widely separated and emerge in a three-dimensional array. Peeking out from their lower edges are their yellow, crescentic flowers. A waxy bloom like silvery talcum powder, typical of many heliconias, is particularly copious on its flowerheads and leaves. Since this stains clothing, do not plant red collinsiana close to walkways.

Although native to Central America, red collinsiana was an early introduction to Hawaii. It is well established in botanical gardens and parks, Iao Valley State Monument, Maui. Its primary blooming period is summer through fall.

Hanging Lobster Claw

Heliconia rostrata

Other Names: firecracker, rostrata, dwarf hanging heliconia, parrot's beak; variety with spiral bracts: *'Twirl'*

Resplendent in tricolored glory, the hanging lobster claw is simply dazzling: its vibrant colors glow — almost iridesce — with an intensity rare even in the tropics.

One of the first heliconias introduced into Hawaii, this beauty is native to Argentina and Peru. Its two-foot-long flowerheads dangle, around eye-level, amidst six-foot-high foliage. A particularly stunning hedge (left) flanks the winding approach to Lyon Arboretum, in Manoa Valley on Oahu, where the effect is so overwhelming that flower lovers should hand the driver's wheel to someone else!

In Latin American markets it is a common sight to see a dozen or so gaudy forest birds strung up for sale. *Rostrata* means beaked. Was the botanist who named this wiggly necklace of bright-hued flowerheads inspired by the strings of toucan and parrot beaks that he saw for sale in Latin American markets?

Of more relevance to today's world, it may be noted that the red and green colors of the hanging lobster claw spell "Christmas" to market specialists. Thus it comes as no surprise that University of Hawaii researchers are attempting to initiate blooming in December, despite its preference for summer.

Above: During summer's Japanese Bon Festival, departed souls are specially honored. At the Paia Mantokuji Mission on Maui, hanging lobster claws form a prominent part of the cemetery decorations.

Right: A close view of hanging lobster claw.

Hanging lobster claw *(Heliconia rostrata)*.

Platystachys

Heliconia platystachys

Other Names: *red-and-yellow pendula, orange hanging lobster claw*

The tapering seven-inch-long bracts of this stunning heliconia, which blends red, yellow, and green, are shaped like huge rose thorns. Uncommon in cultivation and even in its natural rain forest habitats stretching from Guatemala to Colombia, it blooms primarily during summer. The unopened bracts at the end of its flowerhead are shaped like large oat grains, presumably the inspiration for its scientific name *platystachys*, meaning "a flat, spiky ear of grain."

Pendula

Heliconia pendula

Other Names: roseo-pendula, hanging red, revoluta (H. revoluta *is a larger species*)

From Venezuela comes this striking crimson hanging heliconia with curled white flowers and an imposing design that resembles floral birds poised for flight. These angular, open flowerheads measure a foot long. Tropical gardeners will notice how banana flowers are similar to these. To date, pendula is only available in limited quantities during summer.

New Pink

Other Names: *colgantea*

New pink's elongated, curving bracts impart an elegant touch to this uncommon heliconia. Its yellow-tipped, pink flowerheads hang in three dimensions from a clump of "pseudotrunks" and broad leaves that stand six to ten feet tall.

New pink, discovered as a new species in the 1970s and named during the 1980s, hails from humid rain forests of 300 to 2,500 feet in elevation, of Costa Rica and Panama. As Costa Rica boasts a far-sighted conservation ethic, (it currently protects 20 percent of its land, richly endowed with tropical birds, plants, mammals, and insects), there is hope that new pink will continue to flourish in its homeland as well as in its adoptive countries.

In Hawaii, the new pink flower blooms during the summer.

'Sexy Pink'

Other Names: *chartacea, chartasia*

Bedecked in erotic pink and apple green, this flamboyant heliconia epitomizes the tropics. Emerging from its snaky rose-colored axis are up to a dozen bracts edged with fragile membranes, from which the name chartacea, meaning "membranous," is derived. From beneath each bract's lower surface, delicate matching florets peek out shyly (see title page). 'Sexy Pink,' as befits her name, sneaked secretly into Maui as an accidently mislabelled rhizome. A few months after planting, her glorious, two-foot-long pendant delighted flower farmers. Imagine the meticulous care taken to nurture this precious plant and its offspring.

'Sexy Pink' ranges naturally from Guyana to Peru. Flowers and rhizomes of several color forms are available by mail order.

Fuzzy Orange Hanging Heliconia   *Heliconia mutisiana*

Other Names: *metusiana, fuzzy pink hanging heliconia*

This delightful fuzzy heliconia has both its snaky axis and orange bracts completely covered with fine white hairs. It created quite a stir amongst heliconia farmers in the 1980s when it first arrived in Hawaii. Native to Colombia, it grows to ten feet in height, bearing an abundance of foot-long flowerheads during spring and summer. Unfortunately, the blooms soon lose their shape. If left on the bushes, they last much longer, and their florets eventually mature into turquoise pearls of rare beauty.

An assortment of gingers mingles with decorative foliage: (mid-left, clockwise) pine torch, Yamamoto torch, pink torch, pink ginger, red ginger, golden beehive, 'Jungle King' ginger.

GINGERS

Flowery red "footballs," layered ice-cream cones, scaly tropical "snakes," delicate and fragrant "butterflies" . . . many ornamental gingers, primarily from Indo-Malaysia, have been introduced into Hawaii during the past century. Though the word *ginger* conjures up mental images of oriental food, edible ginger is only one of approximately 1,400 species in the highly diverse family of Zingiberaceae.

Red, white, yellow, and kahili gingers, intimately associated with island beautification and decoration for decades, are assumed to be native, even by residents. But, in fact, they are really just another aspect of Hawaii's Asian heritage. Red ginger is, however, native to Polynesia, but it never reached Hawaii on its own. Though gingers are basically tropical, a few gingers are of Himalayan origin. Their cold tolerance enables them to grow on Hawaii's higher slopes as well as in many mainland states, especially if mulched in winter.

In contrast to heliconias, gingers are renowned for their irresistible perfumes, but the price paid for their delicacy of form and sweet aroma is that they do not last quite as long. All but the most fragile gingers are available year-round from flower farms, cooperatives, and florists. If you like the fragile ones — and who doesn't? — it's all right to pick a few yellow and white gingers along the roadsides.

Tropicals are quite different from familiar blooms like roses and daisies. As with heliconias, the most conspicuous parts of each ginger flower are not petals but scale-like bracts or highly modified parts of the male reproductive system. Whoever heard of stamens mimicking fluffy petals?

Instead of "regular" roots, gingers, like heliconias, spread and are propagated by fat, knobby, ground-level stems called rhizomes. When buying fresh so-called ginger root at the market, one actually receives a specialized stem. Many Asian spices, coveted items of trade and commerce since the days of Columbus, are time-honored specialties of the ginger family. For example, turmeric powder is from a dried ginger rhizome. Today, a new array of gingers, equally exciting, appeals to our senses, but this time to our visual and olfactory, rather than taste, senses.

Pink, red and Yamamoto TORCH GINGERS rest in a stream prior to an overseas flight. Soaking them again after the trip extends their lives considerably.

White Ginger

Hedychium coronarium

Other Names: *common ginger lily, butterfly ginger, garland flower,* 'awapuhi ke'oke'o

> *"White ginger blossoms cool and fragrant, sweeter than the rose, fairer than the moonlight,*
> *White ginger blossoms from the mountains fill the thirsty air with exotic fragrance rare."*
>
> — From a nineteenth-century poem by R. Alex Anderson

Cultivated in their native Asia for millennia and in Hawaii for decades, both white ginger and its twin, yellow ginger, beautify Hawaii's windward highways and gardens. Chinese immigrants evidently brought them to Hawaii in the late nineteenth century. Their

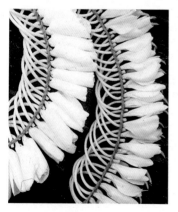

showy, orchid-shaped flowers manifest an ethereal delicacy of form and an enchanting odor. Even their scientific name, *Hedychium* (sweet snow) reflects these attributes. White ginger perfume, though only a pale substitute for the living aroma, is widely available in Hawaii. These gingers are quite cold-tolerant, so if you live in a relatively mild climate with few frosts and warm summers, you may be able to grow them in your yard or in a big tub on the terrace.

These twin gingers are special favorites of Hawaii's *kamaaina* (old-timers), especially for making *lei* (left). Their sweetly romantic perfume lingers for weeks. The blossoms are so brittle they are useless to the floral trade except as garlands, cleverly woven by a method shared by many Pacific islanders. These beautiful *lei* were woven with raffia, Micronesian-style. White ginger has been used for flower head *lei* and garlands for so long that its name, *coronarium* (head-circlet), forevermore celebrates this practice.

Yellow Ginger   *Hedychium flavescens*

Other Names: *yellow ginger lily,* 'awapuhi-melemele *("yellow ginger"), cream ginger*

Even more ubiquitous than white ginger is the irresistible yellow ginger, originally from India. Its pale yellow and gold flowers, blooming prolifically all year except winter, cheer roadsides and gardens throughout Hawaii. Pick a blossom, nip off the bottom one-eighth inch, pretend it is a honeysuckle, and suck a few drops of its sweet, gingery nectar. Although actually intended for its pollinators — long-tongued moths and butterflies — no doubt you will find it delicious. The author once saw a greedy hummingbird moth shove his proboscis so far into a flower that it wouldn't come out! After expending all his energy attempting to fly off, he finally died, dangling pitifully from the spent flower. In Madagascar, essential oils are distilled from yellow ginger flowers to make an expensive perfume.

Flowers, like people, have varied ways of attracting attention to themselves. Here is a species that is decidedly different from the norm. Flowers are usually hermaphrodites, a mixture of both sexes. The most conspicuous parts of this flower are its male reproductive organs. Its three "petals" are in fact stamens, while the three skinny filaments are real petals. The female part is a little green knob on the regular stamen. Recent hybridization, especially in Japan, is producing some beautiful varieties, some of which can be seen at Waimea Arboretum, on Oahu.

Although attractive, yellow gingers, which grow to about seven feet tall, are pests in native forests. Escaping from the confines of home gardens, they creep rapidly along watercourses, spreading their tenacious pink rhizomes in every direction. In some wet forests (for example, above Hana, Maui, and in the Alakai Swamp, Kauai), they have strangled vast stretches of formerly pristine forest. The watercourses winding along some of Maui's most spectacular inland valleys are so choked by gingers they test the endurance of even the hardiest hiker.

Kahili Ginger
Hedychium gardnerianum

Other Names: *yellow India ginger, 'awapuhi-kahili*

From high in the Himalayas comes the king of gingers, the kahili, named for its resemblance to Hawaiian *kahili* or royal standards. These familiar insignia of bygone days — red-and-yellow cylinders made from native bird feathers atop long shiny poles — were indispensable on ceremonial occasions, just as trumpet fanfares and diamond-studded scepters were in Western cultures. Fabric or paper replicas today are displayed on May (Lei) Day and other local celebrations.

Kahili ginger is a stunning ornamental. Each foot-long flowerhead contains dozens of yellow blossoms. Each thrusts out one enormous red stamen for pollination. Like its close kin (pages 44 and 45), the kahili resists packing in boxes as its brittle flowers usually snap off, resulting in a woeful sight.

Its montane origin (up to 8,000 feet elevation in Nepal) predestines it to grow quite happily in cool climates. With winter mulching, it weathers minimum annual temperatures of 10 degrees to 20 degrees F over a large geographic area that includes Seattle, San Francisco and Maryland (USDA Hardiness Zone 8).

RED KAHILI or scarlet ginger lily

46

Red Ginger
Alpinia purpurata

Other Names: 'awapuhi-'ula-'ula *("red ginger"), ostrich plume*

Red ginger's showy flower spikes, composed of loose layers of cerise bracts, are known to all. Prior to the recent heliconia boom, tropical flower arrangements featured primarily red ginger, birds-of-paradise, ti leaves and orchids.

Native to the western Pacific, red ginger has been transported by people to warm areas everywhere. True to its island heritage, it still remains a festive plant. In Hawaii, entire flower spikes are used, whereas in Samoa, where it indicates high rank, individual bracts ("petals") are pulled off and strung into garlands. Blooms, up to a foot long, last two weeks when cut and three weeks on the plant.

Red and pink gingers make excellent hedges or garden dividers. When watered well and provided with plenty of nitrogen, they thrive in sun or shade, blooming all year and occasionally reaching 15 feet high. Cut the blooms early in the morning, leave the stems on them as long as possible, and feed them a little sugar to extend their post-harvest life by at least a week. They grow anywhere in Hawaii and, given adequate mulch, will do well in mainland states where winter temperatures do not fall below 20 degrees F. In states such as Florida, Mississippi, and California, they bloom seasonally.

Above: Red ginger is unusual in that seeds rarely develop. Their primary method of reproduction is by the development of tiny plantlets, aerial off-shoots from inside the red bracts. Each bract becomes a marsupial-like pouch, bearing several plantlets like a kangaroo with quints. The "babies" grow fast, eventually weighing down the mature stem. New seedlings take root in the ground adjacent to the parent plant, and a new generation of red gingers is started.

Left: RED AND PINK GINGERS, twins within the same species, lie in a cleansing bath.

Right: 'TAHITIAN GINGER' (*Alpinia purpurata* cv. 'Tahitian Ginger'), a double-flowered form of red ginger, is a botanical curiosity. Beginning life as a seemingly normal red ginger, each flower spike continues to expand in all directions until it attains the size of a football.

Above: 'JUNGLE QUEEN' AND 'JUNGLE KING' (*Alpinia purpurata* cv. 'Jungle Queen' and cv. 'Jungle King,' respectively), a regal pair, are easily distinguished from the regular pink and red gingers. Flower spikes of the former pair are rounder on top, resembling ice-cream cones, whereas the latter pair are more cylindrical.

Right: PINK GINGER (*Alpinia purpurata* cv. 'Eileen McDonald'), smaller than its red parent at six inches long, came to Hawaii from Tahiti in 1973. Named after a Mauian, this popular cut flower is sometimes planted around hotels and office complexes. Mostly unknown for nine years, it ushered exotic tropical flowers into Hawaii's business world on Labor Day, 1982, at a large plant sale in Hana, Maui. Though typically plain (p. 7) this individual sports a "bloody" streak.

Shell Ginger

Alpinia zerumbet (= speciosum)

Other Names: shell flower, porcelain ginger or "lily," 'awapuhi-luheluhe, *catimbium*

This exquisite ornamental has been cultivated for so long in Hawaii it even boasts an Hawaiian name, *'awapuhi-luheluhe* (drooping ginger). Its foot-long floral necklaces, arching gracefully from tall stems, resemble curved strands of pink-and-white porcelain shells. From beneath the waxy bracts emerge one or two bright red and gold flowers shaped like frilly bells. From a distance the large leaves of shell ginger (five to 12 feet high) look untidy if not well-tended. Shell ginger both beautifies many island gardens and grows wild in windward areas, including along Maui's Hana Highway, the Island of Hawaii's Hamakua coast and Route 11 to Hawaii Volcanoes' National Park, and Kauai's Routes 56 and 50 towards Haena and the Koloa turnoff.

Although associated with the tropics, shell ginger actually evolved in the high-elevation cloud forests of Southeast Asia; it thus grows successfully on Hawaii's mountain slopes and as far north as Oregon and Maine. It tolerates heat and dryness better than most gingers. In its native countries, the fibrous leaf sheaths of shell ginger are converted into ropes. Novelty species include small shell (*A. mutica*), variegated (*A. sanderae*), and upright-flower (*A. calcarata*) gingers.

Torch Gingers   *Etlingera elatior* (= *Nicolaia elatior,Phaeomeria magnifica*)

Other Names: 'awapuhi ko'oko'o *("walking stick ginger"), nicolaia*

Majestically, the torch ginger thrusts its huge flowerheads skyward on thick stalks that the nineteenth-century Hawaiians likened to walking canes. Each heavy flowerhead, up to eight inches in diameter and seemingly artificial in looks, is composed of numerous layers of waxy, cerise frills, neatly margined in white. One of its former names meant magnificent and pure light. Floral scallops, overlapping like exotic pinecones, are not petals but fancy leaves, or bracts. The true flowers are the yellow curly tongues which enlarge as the flower ages. Arising directly from the ground, often up to eye level, torch gingers' heavy blooms are dwarfed by their own enormous leaves.

Native to the island of Mauritius in the Indian Ocean, torch ginger has grown in Hawaii for decades, but because of its enormity is not often seen in gardens. Torch ginger is best accompanied by grandiose scenery, waterfalls and jungly vegetation. Look for it in botanical gardens and along windward roadsides, where you can spot its 20-foot-high leaves in several locations along the Maui's Hana Highway: Kaumahina State Park, Honomanu Bay, Waikani Falls, Keanae, and Wailua Falls, near Kipahulu.

A new pale pink and somewhat spidery torch ginger is the YAMAMOTO SEEDLING, which travels well.

50

Left: Two varieties of TORCH GINGER, red (*E. elatior* cv. 'Red Torch,' left) and the smaller pink (*E. elatior* cv. 'Pink Torch,' right), add drama to large flower arrangements.

Below: A YOUNG TORCH GINGER unfolds.

TULIP TORCH GINGER (*E. hemisphaerica* cv. 'Helani Tulip'), as both common and scientific names imply, is shaped like a combination of a globe cut in half and a tulip. Waimea Arboretum, Oahu, and Helani Gardens, Maui, are excellent locations to encounter this Hawaiian *malihini*, or newcomer.

Golden Beehive Ginger

Zingiber spectabile

Other Names: *golden shampoo ginger, beehive, nodding gingerwort, spectabilis ginger, 'awapuhi ("ginger"), mustard, revolute*

Layers of golden, ridged scallops, intricately molded from waxy golden plant tissue, make up the golden beehive's eight-inch flowerhead. At any one time, two or three spotted, orchid-like flowers peek out between its crescentic folds. The golden beehive's yellow and

red pigments absorb ultraviolet light, attracting bees to pollinate its peculiar flowers. How appropriate that its scientific name means remarkable ginger.

Native to Malaysia, this six-foot-high newcomer to the floral trade is, like its wild relative, quite aggressive, as the prolific clumps of flowerheads attest. If it oversteps its boundaries, dig up the rhizomes and cook with them as the Malays do. Remember that gingers, not as dependent on light as are heliconias, will grow in shady nooks and crannies if given room.

As we all know, the tropics superabound with an astonishing wealth of insect, bird, mammal, and plant species. The center of ginger evolution is Asia, where an incredible 1,200 species occur. Recently, adventurous botanists, trekking off the beaten track in Borneo, Thailand, and India, discovered numerous new species of

gingers — exciting finds in an era when many people assume that all living things are neatly categorized. It is not known how many were inadvertently wiped out during the terrible Vietnam War or are imminently doomed in today's unrelenting destruction of rain forests.

Wild Ginger

Zingiber zerumbet

Other Names: 'awapuhi kuahiwi *("ginger of the uplands"), shampoo or bitter ginger*

Sixteen hundred years ago, precious rhizomes of the original *'awapuhi* (ginger) were carefully stashed in canoes in the Marquesas Islands and transported over 2,000 miles to Hawaii. In traditional Polynesian manner, its leaves were used to flavor meat and fish while they baked in *imu* (underground ovens). In addition, its thick, fragrant rhizomes were sliced, dried, and pulverized into a powder used to perfume folds of stored *tapa* (bark cloth). Fresh rhizomes also relieved a variety of ailments including coughing and indigestion, while green leaves were indispensable in celebrations honoring Laka, the goddess of *hula*.

It is not surprising that the early Polynesians treasured wild ginger, as many members of the ginger family enjoy cultural esteem for their pleasant aroma, pungent spiciness, medicinal properties, or unusual beauty. Do not history books recount adventures and perils encountered by explorers as they sought culinary delights in the "spice islands"? Wild ginger was the only ginger that Polynesians knew, and although it is neither as beautiful nor as tasty as many other species, its rhizomes are edible. Although rather bitter, they are occasionally used for cooking in its native Asia; edible ginger (page 54) is far superior.

Wild ginger's most celebrated use was as a hair cleanser and conditioner, hence its widespread name in Hawaii: shampoo ginger. It is easy to use. Squeeze some of the flowerhead's copious, sudsy juice into a bottle (don't let its sliminess offend you), take it home, and wash your hair or massage a friend's body with it in native style. Bodily perfection was important to the ancient Hawaiian — from birth to old age their bodies were exercised and massaged with wild ginger or oils such as coconut and *kukui.*

Today wild ginger covers many square miles of lush lowland forests. Hikers are familiar with it. A walk along the lushly tropical Tantalus Trail on Oahu, or Waikamoi Ridge Trail on Maui guides you through a sea of this special reminder of old Hawaii. Its large, double-sized variety (above) is exhibited at the Lyon and Waimea Arboretae on Oahu. Cut flowers, purchased through florists and co-ops, keep for two to three weeks. A ground cover in tropical forests in Asia, wild ginger also inhabits the Himalayan foothills and is fairly cold-tolerant. It will grow from California to Washington State, Florida to Maine, and in states bordering the Caribbean.

Edible Ginger   *Zingiber officinale*

Other Names: *common, Canton, commercial, Chinese or Jamaica ginger, ginger pake (pron. "pah-kay") or* 'awapuhi-pake *("Chinese ginger")*

Spring rolls, teriyaki steak, sweet-and-sour pork, stir-fried vegetables and spice cakes . . . scrumptious international treats. They wouldn't be right without fresh ginger. The most valuable "root spice" known, edible ginger is a staple in kitchens from East to West. Once a cook is spoiled by the aromatic pungency of the fresh product, its powdered form becomes only an emergency substitute.

Unfortunately, edible ginger is not as easy to grow as the feral gingers that thrive in Hawaii's lush lowlands. How many people have wished that yellow ginger's gorgeous cerise rhizomes tasted as good as they smelled? When carefully tended, however, yields of edible ginger are high: up to 1,500 pounds of dried rhizomes per acre (below).

Edible ginger's pinkish-white "hands" (left) are cultivated in many tropical locations: Asia, where it is native, Africa, tropical America, and the West Indies. Hawaii exported seven million pounds in 1988, in addition to conducting research into advanced agricultural methods such as tissue culture. Until recently, the highest quality rhizomes were Jamaican, but Hawaii is now providing stiff competition. A superior variety called *mioga* is also cultivated in Japan. Next time you eat at a Japanese restaurant, note the mild, non-fibrous, pink shreds of young ginger that accompany sushi (rice wrapped in paper-thin seaweed). So delicious, and they are hardly recognizable as raw ginger.

Edible ginger (two to four feet high) resembles wild ginger, having fairly upright leaves and loose, plain flowerheads. Gardeners should be aware that some supermarkets treat the rhizomes so they will no longer sprout.

'Olena (Turmeric)

Curcuma domestica

Other Names: *tumeric,* 'olena *ginger, hidden lily, Cape York lily (Australia), white curcuma*

Although carried to Hawaii by seafaring Polynesians around 1,500 years ago and cultivated extensively for centuries, 'olena is today virtually extinct in both the wild and in gardens. Recently it was reintroduced for the flower trade and for cultural education.

Turmeric generally evokes mental images of India and the smells of its diverse curries. Cultivated in Asia for millennia, its bright yellow powder is manufactured from mature, dried rhizomes. Not to miss an opportunity, Indians also transform its young, colorless rhizomes into a powdered thickening similar to cornstarch.

Curiously, Polynesians never made curry sauces, using this prized condiment only for medicines, dyes, and sacred rituals. Today its medicinal properties are not forgotten; a few *ka-maaina* still swear that it relieves earaches and troublesome sinuses. 'Olena's major use in old Hawaii was to make long yellow skirts (*pa'u*) for women. Their satin equivalents are still worn by *pa'u* riders during Kamehameha Day parades. Only the finest white *tapa* was used for *pa'u*, ensuring a bright yellow color when dyed. Three closely-related species are periodically available in Hawaii — the standard *'olena*, with green-and-white flowerheads (below right); rose turmeric, also called curcuma *'olena* (*Curcuma elata*, below left); and the glossy burnt-apricot beauty, Jewel of Burma (*C. roscoeana*, above). All bracts, despite their differences in color, support dainty yellow flowers. Note the similarity of *'olena* to calatheas (page 72).

Costus

From here to the end of this chapter, a group of gingers are discussed that formerly belonged in the costus family (Costaceae) but are now included with true gingers (Zingiberaceae). Possibly the most ancient of all gingers, all are characterized by twisted tissues: spirally arranged bracts on their flowerheads, swirls of leaves up their bamboo-like stalks (10 to 15 feet high), and curling leafbuds.

Indonesian Wax Ginger   *Tapeinochilos ananassae*

Other Names: *Indonesian or Malaysian ginger, giant spiral ginger, pineapple ginger, 'awapuhi Malae ("Malay ginger")*

Another fanciful species, the Indonesian wax is characterized by a pinecone-like flowerhead composed of about a dozen layers of crisp, pointed tongues arranged in a continuous circle. Its colors are vibrant — yellowish-red on the inside merging into rusty-red on the outside. The entire flowerhead, four to eight inches long, looks as if someone sprayed it with clear polyurethane. Each emerges from ground level on a remarkably short stalk. Is this a real plant or a clone of bizarre, multitongued creatures clustered beneath 15-foot-high culms of twisted bamboo?

Indonesian wax, which was introduced to Hawaii in 1959, will thrive in dense shade and is quite a common component of modern arrangements. Recent research in Hawaii has shown that it can be cultivated by cuttings — unusual for gingers — thus by-passing potential problems with soil-borne pests and diseases.

56

Crepe Ginger   *Costus speciosus*

Other Names: *crape ginger, Malay ginger, spiral flag, common spiral ginger,* 'awapuhi

An attractive tall shrub that resembles spiralling bamboo, crepe ginger may be encountered in arboreatae, lowland gardens and lush roadsides in moist regions (Hanalei on Kauai; Hamakua coast on the Island of Hawaii; Hana, Maui). It is not grown commercially, as its large frilly flower is too delicate. Although entirely ornamental in Hawaii, in its native Malaysia, India, and the Philippines, crepe ginger's young shoots are cooked in coconut cream and savored as a vegetable.

The ruffled flowers, emerging one or two at a time, have a curious construction: the large fluted skirt is not a petal but a modified stamen, part of the male reproductive system. The other stamen is centrally located as one might expect (see also yellow ginger, page 45).

People often wonder what is the use of saving individual plants and animals, of preserving ecosystems. One of the multiple reasons is that plants are widely used in medicine and new pharmaceutical discoveries are surfacing daily. Of particular relevance here is the recent discovery of potent steroids in the rhizomes of crepe ginger. These chemicals are currently used for birth control pills in India.

Indianhead Ginger
Costus spicatus

Other Names: *Pink Indian Head, closed spiral ginger, scarlet spiral flag,* 'awapuhi - 'Inikini Po'o *("Indian Head ginger"), costus*

Flaunting a yellow flower projecting from a reddish cylinder — mimicking a single feather on an American Indian headdress — the indianhead ginger is yet another unusual, though somewhat less showy, tropical plant. Its blossom never opens fully: the forces of evolution have ensured that its tubular design is perfect for pollination by hummingbirds. Hawaii has no hummingbirds, and in their absence, the yellow pigments absorb ultraviolet light, enhancing the blossom's attraction for pollinating bees.

Spirally arranged leaves on a twisting stem is a typical growth habit of all 140 members of the genus Costus. This one, native to tropical America, is Hawaii's most familiar. It makes a fast-growing background planting, especially where atmospheric humidity is high, and it can be seen in large estates, botanical gardens, and heliconia farms. Blooming year-round, its two- to four-inch-long flowerheads provide distinctive material for mixed floral arrangements.

In tropical America, indianhead ginger's sap is used as a diuretic.

A close relative from Brazil, ORANGE COSTUS (*Costus scaber*) is occasionally seen in cultivation.

Red Snake

Costus stenophyllus

Other Names: *cobra, snake, stenophyllus*

Bright reptilian flowerheads, looming directly from ground level, mimic a tropical snake complete with overlapping "scales" and the standard warning colors of red and yellow. Never fear, for it is harmless, providing a sure topic of conversation as it lurks behind the scenes in a contemporary flower arrangement.

Its spiralling, bamboo-like leaves are rather narrow, as *stenophyllus* (narrow leaves) implies. Red snake does not require intense light, growing well in shady spots where heliconias do not have a chance. It is available in limited quantities from heliconia farms and flower cooperatives.

Obake anthuriums (pron. "oh-bah-kay"), variable in their subtle combinations of green, white, or pastel colors from the red side of the rainbow, are Hawaiian in origin despite their Japanese name. Obake hybrids first appeared in the 1930s.

MISCELLANEOUS EXOTICS

Miscellaneous exotics include familiar flowers commonly associated with the lush tropics: romantic anthuriums, perky birds-of-paradise, diademed blue gingers, and stately traveler's trees. Dotted in-between are lesser-known gems: glossy calatheas, reptilian rattle shakers, floral cigars, and ornamental pineapples. Most of these flowers are *malihini* (newcomers) — ice-blue calathea is a freshly discovered species — and all are available through the state's mushrooming tropical flower trade. Because of their diverse botanical relationships, they are arranged alphabetically below.

Blazing with fiery color, the brilliant OKINAWA TORCH, an ornamental banana, will last five to six weeks after cutting.

Anthuriums

Although associated today with Hawaii's tropical glamour, anthuriums, like most other landscaping gems, are not native to the islands. The original anthurium was brought to Hawaii by an English missionary, Samuel Damon, in 1889 from Colombia, via London. Denizens of muggy rain forests in tropical America, their numerous species are at home perching on branches amidst the dense shade of multi-layered tiers of lush, dripping vegetation. It is no wonder that their cultivated offspring abhor direct sunlight and dry air.

Anthurium's kindred plants, all in the arum family (Araceae), include taro, caladium, philodendron, and monstera from the tropics, and Jack-in-the-pulpit and calla lilies from more temperate climates. All share a similarly shaped flower: a shiny, heart-shaped "petal" (spathe) surrounding a curved knobby "phallus" (spadix). Reasonably priced, anthuriums last several weeks after cutting and they travel well, thrilling people all over the world with fantasies of Hawaii.

Anthuriums grow best in hothouses with controlled humidity, though some varieties are now designed for indoors. Originally a backyard venture, the cultivation and hybridization within recent years of these almost unreal "plastic" flowers has become commercially profitable to the extent that annual wholesale sales now exceed $8 million. Almost two million dozen are now shipped worldwide each year. Ninety-five percent of flowers come from around 200 farms on the windward side of the Big Island, whose warm, drizzly weather provides ideal growing conditions.

Left: The BRIGHT REDS, ranging in size from tiny (peewees) to extra large, account for almost half of Hawaii's anthurium sales. When they first appeared in Hawaiian markets, in the 1940s, their colors were much paler than those of today. *Right:* OBAKES, though often clothed in muted greens and pinks, are sometimes boldly colored. *Obake* means "ghost" or "change" in Japanese, alluding to their delicate, somewhat nebulous coloration, and also to their unpredictable individual variation. One exhibition obake bloom measures a whopping 12 inches in diameter.

Below: *Midori,* Japanese for "green," is not only the name of a contemporary prodigy violinist, but also applies to beautiful GREEN HYBRID ANTHURIUMS. Anthuriums thrive best between temperatures of 65 and 80 degrees F. Many anthurium cultivars bear Japanese names ('Kansako,' 'Ozaki') or Hawaiian names ('Pahoa,' 'Anuenue').

Above: WILD ANTHURIUMS are encountered abundantly in wet, lowland forests of tropical America. Their flowers range in size from very tiny to about eight inches long. Like wild corn and tomatoes, they are plain, generally bearing green

Left: BUTTERFLY ANTHURIUMS, developed in Hawaii, are obake hybrids with a modified heart shape. They favor shady garden nooks.

Right: On the Island of Hawaii, 500 acres of anthuriums are grown commercially under both natural and artificial shade at one of the state's oldest plantations, Shimabukoro Farm, where anthuriums are grown under *hapu'u* (treeferns).

Right: Enclosures utilize fine-gauge netting similar to window screening. Here, pink anthuriums thrive in a 20-acre "greenhouse" at Puna Flowers and Foliage. Anthurium compost uses volcanic cinders and recycled products from Hawaiian industry: taro peelings, sugar cane fiber (bagasse), treefern chips, wood shavings and macadamia nut shells.

Left: Of basically upright form, 'CALYPSO' is a tulip-type, novelty anthurium. Each spathe is cupped like an individual tulip petal. *Right:* Another OBAKE anthurium, subtly blending pale apricot and green.

Top Left: Hybridization of anthuriums involves hand-pollination, monitoring of seeds on the knobby spadix, and meticulous nurturing for two or three years. ***Top Right:*** Buckets of anthuriums await shipment from Green Point Nurseries. ***Above:*** Glowing with luminescence, this apricot hybrid, which pales with age, is called 'ASAHI' ("sunrise").

65

OKINAWA TORCH (red-flowering banana, red ornamental banana, red Thai banana, or Okinawan banana flower), about eight inches long and six inches in diameter, is from Southeast Asia. Its scientific name, *Musa coccinea,* means "scarlet banana." The orange-red bracts, lightly tipped with yellow, do not curl backwards in typical banana fashion.

Bananas, Ornamental

Musa spp.

Lasting more than four weeks after cutting (three months on the plant), ornamental bananas are initiating excited interest in the field of floral research. Related to the numerous types of edible bananas, these small broad-leaved plants bear stiff, upright buds that can be cut when young and tight, then shipped without fear of bruising. Over the next few weeks their vivid bracts unfurl one at a time, exposing successive layers of fresh yellow flowers, while bunches of mini-bananas develop from the jointed stem.

Asian in origin, ornamental bananas are already adding elegance to a few botanical gardens and parks in wet areas. Akaka Falls State Park on the Island of Hawaii and Lyon Arboretum on Oahu, both bursting with tropical verdure, are good places to see them. Like edible bananas, their entire fleshy trunks are slashed down after flowering. Young shoots, already growing around their bases, then shoot up into the light.

Both edible and ornamental bananas can be grown successfully elsewhere by overwintering indoors. They may not always fruit, but the self-peeling banana has produced flowers as far north as Chicago.

Left: The stunning SELF-PEELING BANANA (*Musa velutina*) is also known as pink-fruited banana. *Right:* One of the smallest species at only six feet high, its conical buds are composed of many pinkish-mauve bracts which unfurl one at a time. The rows of yellow flowers mature into fuzzy fruits (*velutina* means "velvety"), which have the bizarre habit of peeling themselves. As each banana ripens, its reddish skin automatically peels back, exposing a white, seedy pulp. Although the flesh is inedible, its black seeds, unusual in bananas, will germinate if provided with the right conditions.

Bird-of-Paradise

Strelitzia reginae

Other Names: strelitzia, crane's bill, bird, reginae

The perky and boldly curvaceous flowerhead of the bird-of-paradise resembles a long-necked bird with a golden crest. Each flower merges in full color with classic poise and floral self-assurance. What perfection of line! Arising from a boat-shaped basal sheath, up to six dazzling flowers emerge over a period of a week or two in a sunburst of glossy orange and blue. The shape is so novel it is difficult for people to figure out what's what in the way of petals and other flowering parts.

This strange "avian flower," whose scientific name means queenly strelitzia, evolved in Madagascar, a large island east of Africa. It is one of the few flowers in the world pollinated by birds' feet. The unusual shape of the bird-of-paradise evolved simultaneously with the evolution of sunbirds, Africa's equivalent to hummingbirds. Pollination is an ingenious

mechanism that only birds are heavy enough to trigger. Attracted to the orange sepals, a sunbird alights on the blue arrow, a synthesis of petals and reproductive parts, splitting it in half. In doing so, previously hidden pollen becomes stuck to its feet. After sucking a full beak of sweet nectar, the bird flies off, its pollinized feet ready to passively fertilize the next flower. In Hawaii, hand-pollination is necessary for seed development, as neither sunbirds nor hummingbirds occur here, bees are too light, and the only other possible pollinators, native honeycreepers, live primarily *mauka* (towards the mountains) of homesites likely to have birds-of-paradise. The resultant woody seeds are so hard they must be filed or soaked in acid before they will germinate.

White Bird-of-Paradise
Strelitzia nicolai

Other Names: *alba, giant white "blackbird"*

Closely related to the orange bird-of-paradise and occasionally spotted in gardens and parks, the white bird-of-paradise is another stunning ornamental. Its enormous "birds" grow on a plant that has long-stalked, banana-like leaves arranged like an enormous green fan (the plant resembles traveler's tree, bottom photo). Striking blue and white flowers, about three times the size of its smaller cousin, arise from a blue-gray "boat," which may be two feet long. Overall, this flower is less appealing than its orange brother, as the flowers have short stalks, and its bulky plants need frequent pruning. Their dead flowers and accompanying slime can look pretty messy. Bring them inside, though, clean them, and they will contribute to a commanding flower arrangement.

The TRAVELER'S TREE (*Ravenala madagascariensis*), extremely similar to the white bird-of-paradise, is unmistakable (right). Arising from a rounded trunk, a huge, two-dimensional fan towers up to 40 feet high, its constituent leaf-bases overlapping to form a neat woven pattern. Some assert that under natural conditions the traveler's tree's "fan" orients itself north and south, providing a compass direction for the lost traveler. Others claim that voyagers can quench their thirst from the ample water held between the leaf-bases. However, judging by the tree's tendency to collect large amounts of debris, and the conglomeration of wrigglers, frogs, etc. that inhabit these aerial ponds, better water could probably be found elsewhere.

Traveler's trees, originally from Madagascar, thrive in warm, wet areas. Watch for them in botanical gardens, the Royal Hawaiian Shopping Center in Waikiki, Wailea and Hana on Maui, and in Hilo on the Island of Hawaii. Although often called traveler's palm, they are unrelated to true palms.

Bromeliads or Ornamental Pineapples

Delectably sweet, a sun-ripened Hawaiian pineapple is a taste treat hard to beat. However, most people become so preoccupied with this renowned fruit that its beautiful, multilayered flowerhead is overlooked. A small conceptual jump from Hawaii's most famous fruit, the pineapple, leads us to its fancy relatives, ornamental pineapples and other bromeliads. All are members of the large family Bromeliaceae, whose approximately 1,500 species are native to the American tropics. None evolved in Hawaii. Today, bromeliad culture is a rapidly-expanding horticultural science. Their beauty can be seen in many public places — hotels, botanic gardens, airports (for example, the Los Angeles Delta terminal). They also are becoming popular both as indoor and patio decorations. Most are recognizable by their rosettes of stiff or glossy leaves, but a notable exception is the familiar "air plant," Spanish moss.

Above: Dripping with sticky sweetness, luscious PINEAPPLES (*Ananas comosus*) ripen on Maui. All fields are commercial and private: take only pictures. If you find yourself behind a yellow pineapple truck on the highway, inhale deeply of its honey-sweet aroma.

Above right: purple flowers emerge from the pink bracts that comprise the complex flowerhead. The entire mass eventually coalesces to form the familiar fruit. This is a new variety (*Ananas bracteatus* var. *Striatus*) with variegated leaves, whose white pulp contains less acid than standard pineapples.

Right: In cultivation, bromeliads thrive in pots of treefern fiber. You can also attach them to sticks, around which they wrap their roots like orchids, or suspend them in air by fishing lines. Pictured is 'SCARLET STAR' (*Guzmania lingulata*) from a hotel lobby on Maui.

70

Bromeliads, though extremely well represented in both numbers of species and individuals in tropical American rain forests, are also common inhabitants of dry forests. Above is a "PAINTED FEATHER' (**left)** and 'DANCING BULB' (**right)** (*Tillandsia bulbosa*) amidst Hawaii's *aa* lava. Under natural conditions, most would be clinging to tree trunks, exposed to scorching heat for part of the year and pelting rains for the rest. Flower-fanciers thus can enjoy the ability of bromeliads to withstand wide ranges of temperature and humidity. Drought-tolerant species possess stiff, whitish leaves and other adaptations to conserving water.

Above: Many bromeliads are epiphytes or perching plants, which do not require soil. As they derive no food — only lodging — from their hosts, they are not considered parasites. Note the relatively wide, soft leaves of this rain forest beauty.

Right: In a capricious mood and armed with a palette of gay colors, nature painted the flowerhead of this ORNAMENTAL BROMELIAD *(Aechmea fasciata)* a rich rosy pink, dusting it with a powdery frosting.

71

Calatheas

Seeing is believing, but seeing is questioning, too. Can Nature mold flowers from unblemished glacial ice? Although introduced to Hawaii in 1973, these three other-worldly Brazilian flowers went unnoticed until they suddenly appeared on the world flower market in the mid-eighties, as though they had materialized from the spiritual ether.

Calatheas (pronounced *kall-ah-thay-uh*), of which there are about 150 species (Family Marantaceae), are valued for their bright-veined, variegated, and shapely foliage. Most dot the dim rain forest floors of the South American tropics. As remote areas become increasingly accessible, more species are discovered and propagated; unfortunately, *each year an alarming 20 million acres of tropical rain forests are entirely obliterated.*

ICE-BLUE CALATHEA, or blue ice, is a newly-described species (*Calathea burle-marxii* cv. Ice Blue') of ethereal beauty. The delicate translucency of its glassy blue tissues and mauve-crowned white flowers, the sensuously-curved network of its tiny veins, and the gentle undulations of its bract margins must be experienced first-hand to be truly appreciated. It is difficult for sellers of tropical flowers to fulfill every order for this crystalline jewel, pictured above with green and white ices.

Above: WHITE CALATHEA, or white ice, is a form of ice-blue calathea that develops from certain seedlings. Its delicate texture does not allow it to travel well.

Left: ROSE CALATHEA (*C. warscewiczii*) radiates a rich cream hue delicately margined in pink. The layered, conical flowerheads match its common and scientific name, *calathea*, a Greek word meaning "vase-shaped basket."

CALATHEA WHEAT, strawflower, or fingers (*Pleiostachya pruinosa* = *Ishnosiphon pruinosus*) is an uncommon, grain-like relative of arrowroot.* Before cornstarch became widely available, the rhizomes of many species of calathea in Asia and the Pacific were dried, powdered and mixed with water to thicken foods. The bi-colored leaves of calathea wheat, green with maroon undersides, resemble small heliconias. Like rattle shaker, its dried flowerheads last indefinitely.

* *The Polynesian arrowroot (pia) is from another family, Taccaceae.*

The CIGAR FLOWER (*Calathea lutea*) provides shape and texture in a vase of tropicals. Although its branching fingers are only a few inches long, the foliage is large and banana-like. It is available in Hawaii during winter and spring.

Ginger, Blue

Common names frequently fool the unwary. Amidst a bounty of true gingers, who would have guessed that this beauty is merely a cousin to the spider plant, a familiar hanging house plant, and the dreaded *honohono* grass, a curse to Hawaiian gardeners and sugar planters?

This dazzling Brazilian ornamental, which is neither blue nor a ginger, is an old-time favorite in island landscaping. It is particularly prevalent in long-established, lush residential areas such as Manoa Valley and Hana. Rambling beneath tall trees, blue gingers seem to condense sunshine into glittering jewels — amethyst, perhaps, or violet quartz. Did this magic beauty inspire its scientific names *thyrsiflora* (flowers in a wand) and *reginae* (kingly)?

Blue ginger is not easily available as a cut flower on the commercial market, but given plenty of moisture, it will last for weeks on the bush. Pictured is a particularly lovely magenta species, similar to the common *D. thyrsiflora*, whose flowers are bluish-purple.

Blue ginger's close relatives, spiderworts (*Tradescantia* spp.), have recently been found to be supersensitive to pesticide levels, automobile exhaust, and ionizing radiation. While evidence of harmful radiation may take years to show up in human populations, spiderworts turn from blue to pink in a few days.

Rattle shakers needs pampering with good soil, protection from wind and sun, regular fertilizing, and plenty of water.

Rattle Shaker

Calathea crotalifera *(= insignis)*

Other Names: *shakers, rattle, rattlesnake plant, hirsuta shakers*

Bearing a striking resemblance to a rattlesnake's tail, this botanical curiosity, which strangely is closely related to ice-blue calathea, comes to us from rain forests stretching from Mexico to Ecuador. The rattle shaker's flowerhead, five to six inches long, is plainly two-dimensional and thinner than your little finger. Its serpent-like bracts interweave in a regular pattern. Yellow flowers form interrupted lines at the outer edges.

The rattle shaker's leathery texture enables it to keep well. Popular items for years in Latin American markets, they are now becoming increasingly fashionable, both fresh and dried, in the United States. True to all the common and scientific names applied to it (*crotalifera* means rattle-bearing), the rattlesnake plant's dried seedpods rattle like its annoyed reptilian namesake.

Good places to observe rattle shaker plants are Foster Gardens and Waimea Falls Park, Oahu, and the trail head at Akaka Falls State Park, Island of Hawaii.

A handsome tropical bouquet (center left to right): red *ti* leaf; Song of India dracaena; ornamental pineapple; rainbow heliconia; fan palm; variegated *lauhala*; red, pink, and tulip torch gingers; ornamental pineapple; dark red *ti* leaf; Song of India and yellow-center dracaenas; green *ti* leaf.

TROPICAL FOLIAGE

Green and blue, according to those of artistic and spiritual awareness, are the most soothing of all colors. Little wonder, then, that forested reserves, well-tended gardens, or the cheery company of cut flowers and foliage uplifts our spirits.

Embracing more than mere botany, the sampling of beautiful foliage covered in the following pages incorporates centuries of cultural associations: practical, historical, and mythological. Wherever possible, extant Hawaiian customs have been included that you will invariably encounter. A few of the most special leaves, gleaned from the world's bountiful rain forests and coastlines, have undergone decades of horticultural manipulation, or in the case of Caladium, centuries. They are now even more beautiful than their ancestors. On account of their diverse relationships, the plants in this chapter are arranged alphabetically.

In temperate climates, plants generally bear small leaves. However, in tropical rain forests, at least 80 percent of all species and 80 percent of all leaves are of medium to large size.

Whether long and skinny, round or heart-shaped, judiciously-placed leaves can add the final polish to a flower arrangement. The elegant shapes and variegation patterns of tropical foliage complement the sweeping contemporary lines and spatial balance of today's exceptional variety of exotic flowers.

Left to Right: This simple, yet dramatic, arrangement uses golden beehive ginger nestled amongst ornamental taro, variegated *ti*-leaf and Song of India dracaena.

Caladium

Caladium bicolor

Closely related to taro and anthuriums, caladium is a familiar heart-shaped leaf, not only in Hawaii's florist trade, but in garden shops throughout the mainland. Dating back to 1773, caladium is perhaps the first tropical American plant ever cultivated in Europe; Beethoven was only three years old when English plant-fanciers were proudly tending caladium in their new, steam-heated hothouses. Since the early days of horticulture, scores of caladium hybrids have become available, exhibiting a multitude of patterns in vivid pink, cream, and green. Caladium's somewhat velvety leaf surface did not happen accidentally. Microscopic bumps, similar to our upper tongue surface, assist in light absorption and allow water droplets to spread out in a thin film and evaporate before the next rain comes. This is very important for non-leathery leaves growing in an environment where daily deluges total 150 or more inches of rain a year.

Climbing Pandanus

Freycinetia multiflora

Shaped to perfection, orange fruits of the climbing pandanus nestle within spiral tufts of leaves. From Southeast Asia to Hawaii, many species of sprawling, palm-like vines like this one scramble up tree trunks to blanket large expanses of understory verdure. Throughout their geographic range, pandanus leaves and aerial roots were traditionally woven into mats, baskets, fish traps, sandals, etc. The vine's tender new growth, rich in vitamins and minerals, also provided a valuable tonic. Its name, pandanus, is one of the few words in the English language derived from Malay: _pandang_ means screw pine.

Above, above left: Sometimes called ornamental freycinetia (pron. "fray-sin-et-ee-ah"), the CLIMBING PANDANUS will last two or more weeks after cutting. Due to its virtual lack of flower stem, a common feature in rain forest vines, the flower and its attached leaves may be woven into a headband. Look for one if you ever attend a party in Hana, Maui. Its bright bracts — splashes of color punctuating forest greenery — have been prized as ornaments and integrated into mythology for centuries.

Left: Hawaii's native '_IE'IE_ (_F. arborea_) symbolized a child who loved trees so much that the gods turned her into a fiery-eyed forest damsel, forever to entwine with her lovers.

Dieffenbachia

Dieffenbachia picta cv. 'Amoena'

Although many people shy away from scientific names, claiming they are too complex to remember, a surprising number have sneaked into the English language: bougainvillea, heliconia, hibiscus, protea, anthurium, philodendron, and dieffenbachia. Many tropical plants have no alternative name.

In the case of _Dieffenbachia_, the other choice is "dumb cane." The name is not as silly as it sounds. In the West Indies last century, one of the slave-master's favorite tortures was forcing imported Africans to chew the leaves of a robust dieffenbachia vine (_D. seguine_). Its sap, loaded with calcium oxalate crystals, was so irritating it rendered the slaves speechless — dumb — for several days at a time. Not surprisingly then, _Dieffenbachia_ sap has been used as a local anesthetic in native medicines.

Dieffenbachia was named after a German botanist and physician, J. F. Dieffenbach, whose productive life in the early 1800s centered around tropical exotics, then a fad among European noblemen. As a result of research and hybridization by him and many others, dieffenbachias are today inexpensive house plants everywhere. In Hawaii, they fare equally well indoors or outdoors, where they require shade. Characterized by thick stems reinforced with woody "rings" and large variegated leaves, they originally hailed from tropical America.

Dracaenas

During the eighteenth century, Italian violin makers prized the dragon tree, a multi-branched giant dracaena, not for its foliage but for the thick blood-like sap that varnished their precious instruments. Today dracaenas are valued for tropical, palm-like elegance.

Members of the large lily family and closely related to yuccas, aloes, and *ti*, the 150-odd species of dracaenas are native chiefly to Asia, Australasia, and Africa. Hawaii's most common native dracaena, *halapepe (P. aurea)*, which resembles a tree-like yucca, is best seen near Kokee, Kauai. Entire tufts and individual strands of dracaenas are versatile additions to flower arrangements. Below left is a DRACAENA BOUQUET: (top left to right) grandiflora, Song of India, Warneckii, and centrally, compacta.

Above: Leaves of YELLOW CENTER DRACAENA (*P. fragrans* cv. 'Massangeana'), broadly banded with yellow, average two to three feet long and four inches wide. A common ornamental, its beauty develops when shade and humidity are high. Stems of this and other dracaenas sprout easily in water. This African dracaena is also called variegated dragon-lily, cornstalk plant or massangeana.

Left: COMMON DRACAENA, *P. concinna* (= *Dracaena marginata*) is a popular ornamental, both indoors and out. Its tufted clumps of narrow, plain green but pink-edged leaves atop long stalks were first planted in Hawaii during the 1920s at the old Bishop Bank in Hilo. Thereafter, an association with banks gave it the nickname "money tree," with its corollary of bringing good luck. Common dracaena, native to the island of Mauritius (Indian Ocean), is the parent of 'TRICOLOR' DRACAENA (*P. concinna* cv. 'Tricolor'), a Japanese cultivar developed from mutations resulting from the Hiroshima atomic bomb. Its leaves sport cream and pale-green stripes suffused with glowing pink.

Hala

Pandanus odoratissimus

Other Names: *pandanus,* lauhala *("leaf of the hala tree")*

Hawaiian mythology relates how this handsome tree, with its spirally-tufted leaves, stilt-like roots, and pineapple-like fruits, contributed to the origin of mankind. Once a beautiful goddess was busy cutting strips of *hala* leaves for mat-weaving when the trimmer made from sharp shell slipped and cut her finger. The resulting pool of blood coagulated and two eggs formed. From these eggs emerged the mother and father of the human race.

Brought to Hawaii centuries ago by Polynesian seafarers, the picturesque *hala* had been revered, utilized and loved for centuries by Pacific people. It has been an integral part of oceanic survival, helping pioneers become established in unknown lands. Even on tiny atolls too dry for coconut palms, *hala* survives; its hardy, wave-tossed seeds will even germinate on bare lava and burning sand. Every part of the tree provides materials for housing, food, medicine, ornaments, fishing, religion and folklore. Today it graces coastlines, gardens, and office buildings throughout Hawaii. Along with coconut palms, it is a true vestige of former civilizations that depended heavily on the bounties of land and sea. It still contributes to daily life on islands far removed from regular shipping routes. Because of this, *hala* was featured in survival courses on Oahu for GIs in the 1940s. Part of their training was as an island castaway for two days, surviving only on rain water, coconuts, fish, *hala,* and other native Pacific plants.

Perhaps, in part, because of these strong cultural associations, the leaves of *hala* and vari-*lauhala* (*Pandanus veitchii,* below left), and *ti* (p. 94) are the most frequently requested foliage greens in the flower trade. A word of warning: If you receive a shipment of *hala* containing a foot-long cluster of tiny fragrant flowers surrounded by white bracts, beware! This is the famous hinano of Hawaiian poetry, reputedly a powerful aphrodisiac. Local maidens lured young men either by chasing them with dangling flower clusters, or by adding *hala* pollen to coconut oil. They rubbed the fragrant oil on their bodies or casually sprinkled it on sleeping sheets of bark cloth.

Female *hala* fruits — green, maturing to orange (below right) — are commonly seen along Hawaii's highways and secondary roads in coastal areas. Locals sometimes jokingly refer to them as "tourist pineapples." To receive a *lei* of individual orange fruits, symbolic of a new beginning, is still a treasured Hawaiian tradition.

HALA LEAVES, black lava and sparkling seas: this special combination of tropical beauty is commonplace on all high volcanic islands in the Pacific. *Left:* a remote area of Maui's north coast near Hana. *Below:* sun-dried leaves in Savaii (Western Samoa).

Above: rolls of de-spined leaves on coconut matting ready for weaving (Christmas Is., Kiribati). Since Hawaii's recent modernization, *hala* mats are no longer woven in the islands, but old ones may be seen in churches and museums. However, the art of *hala* weaving is not dead: many Pacific islanders not only still weave and sleep on *hala* mats but export their woven handicrafts to Hawaii.

Laua'e Fern

Microsorium scolopendria (= *Polypodium scolopendria*)

In early Hawaii, it was impolite to mention the name of a sweetheart directly, so she was veiled in, and honored by, the name of a plant such as *laua'e* fern.

Bearing fronds that resemble breadfruit leaves, the hardy *laua'e* is one of our most

attractive and well-known ferns, beautifying Hawaii's windward roads and forming verdant nooks at resorts. (Though mentioned in Hawaiian folklore, curiously botanists did not collect any until the 1920s.) Its snaky green stems intertwine at ground level. Carpets of their shiny leaves, indented with finger-like lobes, are indeed an agreeable sight. Consequently, ferns have traditionally been used in the islands for adornment and decoration, as in the modern dancer below wearing a head *lei* of young *laua'e* and a rope *lei* from *ti* leaf. To share that grandest of all emotions, love, a bouquet of anthuriums and *laua'e* cannot fail to deliver a fond message.

Where growing conditions are optimal — plenty of water, warmth, and shade — mature leaves, normally about one inch long, may exceed four feet. Note how the globular spore-capsules depress the fronds so deeply that their warty impressions can be seen from the top side. When cut, fronds last about a week.

A similar wild species, also common in Hawaii's lowlands and lower mountain slopes, is *laua'e haole* (foreign *laua'e*, *Phlebodium aureum*), which is easily distinguished from the regular *laua'e* by its bluish-white leaves and rusty, scaly rhizome.

Peacock Plant   *Calathea makoyana*

Each leaf of the peacock plant is an artistic masterpiece: curvaceous designs of opaque tissue overlap a pale, paper-thin background to create a leaf-within-a-leaf pattern. To accentuate its beauty, the arching motif is deep red below. When light shines through the leaf, a scintillating blend of translucent and bold colors greets the eye: pink, silver, ivory, reddish-purple, chartreuse and deep green.

A popular house plant for many years, this denizen of gloomy Amazon rain forest floors is even more alluring when grown under conditions that closely approach its natural habitat — high humidity, heavy shade, and moist rich soils. Its variegated foliage is a commonly-requested item in the tropical flower trade. How many undiscovered beauties hide in the world's equatorial rain forests, which disappear by a staggering 50 acres each minute?

The peacock plant is closely related to prayer plants (*Maranta*), blue ice (page 72), and *Calathea glazioui* (right).

Monstera

Other Names: *cut-leaf philodendron, swiss cheese plant, fruit salad plant, window plant, ceriman, Mexican breadfruit*

Monstera's huge, multi-fingered, holey leaves, up to three feet across, are unsurpassed for creating a tropical ambience. Whether artfully arranged or spreading their charm in island gardens, they demand attention. A Central American native, monstera's kin include anthuriums, calla lilies, and house plants such as philodendron and taro. Calcium oxalate, a highly irritating chemical, and mildly toxic proteins suffuse their tissues. However, when soft and very ripe, monstera's cylindrical fruits are perfectly edible, though too prickly-like for some people. The flavor is unique, tasting like a delectable mixture of bananas, pineapples and soursops. A charming description of this six-sided segmented fruit is from a Mr. Dobrizhoffer, who chanced upon it in a steaming Mexican jungle in 1784: "Its liquid pulp has a very sweet taste but is full of tender thorns, perceivable by the palate only, not by the eye, by which account it must be slowly chewed but quickly swallowed."

Monstera's perforated, leathery leaves appear particularly gigantic indoors; they need large vases and heavy metal "frogs" to stabilize them. Even "tiny" leaves are ten inches across! As a living ornament it is easily tamed, growing best in a large pot against a support that retains moisture, such as a mossy totem pole. The muted light intensities of an office or living room — surrogates of its natural habitat — simulate the shadowy forest floor beneath a dense canopy.

Up to three feet across, monstera leaves create an instant tropical ambience, indoors or outdoors. It is no surprise that one of monstera's nicknames is SWISS CHEESE PLANT.

The grandeur of Rainbow Falls, near Hilo, on the Big Island, is enhanced by monstera's rambling foliage.

When soft and ripe, monstera's fruits are edible, tasting like a mixture of bananas, pineapples and soursops.

A particularly lush MONSTERA embellishes a palm trunk. Monstera always has holey leaves; if a monstera-like vine does not, it is probably a split-leaf philodendron (*Philodendron radiatum*).

A contemporary floral design focused on 'Sexy Pink' heliconia and embellished by rattle shaker, parrot's beak and other heliconias, Yamamoto torch ginger, and monstera. *Courtesy of Hana Tropicals.*

Philodendrons

Many visitors to Hawaii expect the islands to be replete with lush jungle scenes: heart-leafed vines smothering forest verdure, naked nymphs showering in roadside waterfalls, and huge orchids dangling from trees. This is only partly true.

Hawaii's native plants, 92 percent of which are found nowhere else on earth, and many of which are beautiful in their own right, do not look anything like vegetation in the continental tropics. However, many tropical ornamentals have been introduced, and where these abound, either tended or escaped into the wild, Hawaii often does resemble classic rain forests. The vines and fancy-leaved foliage contributing to this luxuriance include dozens of philodendrons and palms, mostly from tropical America. Those available commercially are perfect complements to similarly exotic flowers such as heliconias, orchids, and anthuriums.

Loosely translated, philodendron means "loving to climb on trees." All 240 species of philodendrons are tropical-looking vines bearing heart-shaped or strap-like leaves. Some, such as split-leaf philodendron, resemble monstera; others have borders embellished with fingers, lobes or scallops. Their leaf-textures are thick and leathery, glossy or velvety.

Like most of its kin, 'EMERALD QUEEN,' a hybrid of two unknown species, prefers shade, high humidity and warmth. It is a squatter rather than a climber.

In many verdant, semi-shaded areas of Hawaii, POTHOS vines (*Epipremnum pinnatum* cv. *'Aureum'* = *Scindapsus* or =*Pothos aureus*) have escaped from cultivation to scramble up trees and blanket the ground, forming jungly curtains of heart-shaped leaves. **Below left:** Introduced from the Solomon Islands, they are especially noticeable near the beginning of Kalalau Trail on Kauai, in photo, Waikamoi area on Maui, and around resorts. **Below right:** A riotous array of philodendrons and monstera flank a path to the 442-foot-high Akaka Falls, on the Big Island.

The shiny, taro-like leaves of BLUSHING PHILODENDRON (*P. erubescens*) from Colombia beautifies many gardens in Hawaii. Its leaves emerge bronzy-red, armed with sunscreen-like chemicals enabling it to tolerate more sun than is usual for this plant group. A feature of rain forest leaves is a long-pointed drip-tip, seen here, which allows water to trickle off the leaf at just the right angle. This not only rids the leaf of excess water but prevents growth of algae and fungi.

Taro

Taro, cherished cargo in early canoes bound for Hawaii and the staple starch when the first European voyagers arrived in 1778, is still grown and savored, mostly as *poi*, by local Hawaiians. To make *poi*, the taro root is peeled, boiled, mashed, and fermented; 80 pounds of raw taro will yield 60 pounds of *poi*. Taro's heart-shaped leaves lend a picturesque,

timeless Polynesian quality to cultivated farmlands such as Keanae/Wailua, Maui; Hanalei, Kauai (above); and Waipio Valley on the Big Island. The Hanalei fields, largest in the state, are managed as a U.S. National Wildlife Refuge, providing federal land for farmers, public education programs, and protection for endangered water birds. With the Hanalei River headwaters arising from 5,148-foot Mt. Waialeale — the wettest spot on earth with 465 inches of rain a year — it is no wonder that the latter's fields are so productive.

During taro's relatively brief 1,500-year sojourn in the islands, it evolved naturally into more than 300 varieties. Black taro (*kalo 'ele'ele*, left) is still prized. The familiar Hawaiian word *'ohana* (extended family) derives from the close cluster of tubers that surround the parent "root," an etymological example of the spiritual closeness of Hawaiians and their staff of life, *poi*. In the olden days, *tapa* (bark cloth), dyed black by burying in black taro mud, was used for burying the dead. Regarded not only as the progenitor of the human race, but also as the actual embodiment of Kane, the Hawaiian's most powerful god and source of all life, taro represented fertility, long life, and the fulfillment of all hopes and desires. From a tiny taro *keiki* (baby) growing in the muddy earth sprang humanity's precious life.

Left: The variegated GIANT ALOCASIA, *Alocasia macrorrhiza*, found in a Nicaraguan rain forest in 1975, has only been in Hawaii a few years. The acrid sap from alocasias counteracts stings from other irritating plants such as nettles. As Central American rain forests are rich in taro-like plants, their medicinal value alone is worth saving. Taro leaves are so waterproof that water just glides off them as though they were plastic-coated.

Note: Ornamental taros are less hardy than regular taros, requiring constant high humidity, abundant water, peaty soils, and protection from wind and sun.

The GIANT "ELEPHANT'S EARS" (*Alocasia macrorrhiza* or *'ape*), pron. "ah-pay," not "ape," is quite common in moist areas and was once used as a love potion. Native to Southeast Asia, it is barely edible, due to large amounts of irritating calcium oxalate crystals in its tissues. **Left:** This plant, a giant even for *ape*, dwarfs photographer Jacob Mau. **Right:** The author's husband pauses by WILD TARO, a frequent dweller of watercourses. Rich in Vitamins A, B-complex, and K, wild taro's tender leafy greens are nutritious and tasty. *Luau,* the name for taro greens, also became the name at some point in history for a Hawaiian feast, becoming one of the few English words derived from a Polynesian language. Incidentally, most people like cooked taro greens better than *poi.* Don't eat them raw!

Huki ika iki i ke ki, e kuu pokii la ola . . . Pull hard at the ti root, my brother, and you will live...be strong and courageous.

Ti, pronounced "tea" but unrelated to cups of tea, is unquestionably Hawaii's favorite foliage plant. A distinctive, large-leaved, rosetted shrub and a member of the lily family, *ti* has been, and still is, valued throughout the Pacific. Its multitudinous uses include food wrappings, plates, sandals, raincoats, medicines, fishing accessories, thatching, toboggans, sweetish confections, "beer," and fly whisks. Try an old headache remedy: wet a *ti* leaf and place on your forehead. *Ti's* shiny, waxy leaves, about two feet long and seven inches wide, are still supplied to florists and restaurants by rural farmers. One widespread use is for *laulau* (rhymes with "brown cow"), an indispensable item at any *luau*: chunks of meat, wrapped in *ti* leaves, are steamed in their own juices in a manner similar to corn-husk tamales.

Hawaii literally bristles with *ti* plants, especially in private gardens. Planted to the right of one's door, *ti* reputedly wards off evil spirits. Although its efficacy cannot be guaranteed, our front door on Maui sports a *ti* on the right and a Japanese good luck plant (*Nandina*) on the left, just in case. If you like this Hawaiian custom, slice off a piece of stem about two feet long, with or without attached leaves, and root it in water or just stick it in the ground.

A staggering array of *ti* varieties now include dwarf, red-margined, and round-leaved types (below left). Cultivar names such as 'Hilo Rainbow,' 'Manoa Beauty,' 'Madame Pele,' and 'Queen Emma' leave little doubt as to their origin.

Although Hawaii has been Christianized for over 150 years, the presence of *ti* leaves in certain places at certain times brings to light pre-missionary beliefs; for example, offerings to Pele, goddess of fire, at Halemaumau Crater on the Big Island.

Above: a *pa'u* rider representing Niihau proudly rides a horse bedecked with small *ti* leaves, native *palapalai* fern, and baby's breath, symbolic of the island's famous shells. **Left:** *ti*-leaf "rope lei" are held by a Hawaiian woman wearing clothes of locally dyed materials. Saffron yellow — a symbol of royalty — is an appropriate partner for *ti,* as its leaves formerly signified a truce between warring chiefs. When *ti* leaves are twisted into ropes, their natural oils lubricate the process, aiding also in their preservation.

ENJOYING TROPICALS:
WHERE TO SEE AND HOW TO BUY

Where to See

Observant eyes will spot tropicals frequently in Hawaii: in commercial and private landscaping, botanical gardens, flower farms, wild along roadsides, and in flower arrangements at resorts, banks, airports, and other public places. Some specific locations for wild lobster claws, yellow gingers, etc. are noted in the individual species accounts. For those who dream of the lush tropics, don't miss visits to the following parks and gardens and to an anthurium or tropical flower farm. The Island of Hawaii (Big Island) is best prepared for visitors. Call ahead to private gardens for business hours (area code 808).

OAHU:
Foster Botanic Garden (531-1939)
Lyon Arboretum (988-7378)
Paradise Park (988-2141)
Waimea Arboretum (638-8511)

MAUI:
Helani Gardens, Hana (248-8975)
Iao Valley State Monument
Keanae Arboretum
Tropical Gardens of Maui

ISLAND OF HAWAII:
Hawaii Tropical Botanical Garden (964-5233)

KAUAI:
Olu Pua Garden, en route Waimea Canyon (332-8182)
Pacific Tropical Botanical Garden (332-7361)

How to Buy

Buying is easy: simply call or write the most appropriate outlet (have your charge card handy), and the order can be delivered to your hotel for hand carrying onto the plane, or else sent airmail to national or international destinations. All plant materials sent from farms have undergone rigorous inspection to maximize quality, are free of diseases, fungi and insects, and will easily pass agricultural inspection. *Rhizomes from friends' gardens or roadsides are not disease-free.* Air freight time to the east coast is three days, and two days to the west coast. When ordering, indicate the size stems you want (up to five feet long); it will save unnecessary expense. Not all species bloom year-round; see species accounts for seasonal availability.

It is impossible to list every outlet for tropical flowers and foliage in Hawaii. The following is a representative sampling of growers and/or shippers who air freight the 136 species and cultivars described in this book. Outlets are retail unless otherwise noted. *An asterisk * indicates specialists in heliconias and gingers.*

*ALII GARDENS: Hana, Maui (248-8288, 248-7218) Major grower and shipper, wholesale. All tropicals except anthuriums.

ANTHURIUMS OF HAWAII: Hilo, Is. of Hawaii (959-8717) Also anthurium plants and orchids.

FLORAL RESOURCES, HAWAII, INC.: Is. of Hawaii (959-5851) Also orchids, anthuriums; wholesale & retail.

FLOWER FARMS, INC.: Oahu (237-8488, 237-8663) Also leis, wedding bouquets.

HANA TROPICALS: Hana, Maui (248-8256, FAX 808-248-7253) Also proteas, anthuriums, and orchids. Located at Hana Ranch.

HAWAII ANTHURIUM GROWERS COOPERATIVE: Hilo, Is. of Hawaii (935-6681, Telex WUI 633127; California (209) 435-8434) Also orchids. Represents 80 member-growers, 35% of Hawaii's anthurium farms.

*HAWAII PROTEA COOPERATIVE: Kula, Maui (878-2525, toll-free 800-367-8047 ext. 215, FAX 808-878-2704) En route to Haleakala Crater; also proteas.

*HAWAII TROPICAL FLOWERS, INC.: Kahului, Maui (871-9578, FAX 808-871-5979) Also proteas, orchids and anthuriums. Near Kahului airport.

HAWAIIAN EXOTICS: Pahoa, Is. of Hawaii (966-9456) Also anthuriums and orchids.

HAWAIIAN FLOWER EXPORTS, INC.: Is. of Hawaii (968-6174) Also anthuriums.

*HAWAIIAN PARADISE NURSERY: Oahu (239-6602, Telex 743-1745) Also ornamental pineapples (bromeliads). Wholesale and retail.

*HUA AINA, INC.: Hilo, Is. of Hawaii (959-9494) Also rhizomes, seeds, plants, dried flowers, ornamental banana plants.

ISLAND TROPICALS: Hilo, Is. of Hawaii (871-8365, toll-free 800-367-5155) Also anthuriums & orchids. Specialize in self-assembling, pre-cut arrangements with foliage.

*KUAOLA FARMS, LTD: Hilo, Is. of Hawaii (959-4565, Telex 633124) Also anthuriums, orchids, rhizomes & plants; wholesale & retail.

*PARADISE PARK: Manoa Valley, Oahu (988-2141).

PROTEA GARDENS OF MAUI: (878-6048, toll-free 800-367-804 ext. 410) Also proteas.

QUINTEL FARMS: Mountain View, Is. of Hawaii (968-8455).

*RAINBOW TROPICALS, INC.: Hilo, Is. of Hawaii (959-4565, Telex 633124) Also anthuriums.

*SUNSHINE FARMS: Mountain View, Is. of Hawaii (968-6312) En route Hawaii National Park. Also rhizomes, seeds, plantlets; wholesale & retail.

*THE TROPICAL CONNECTION: Pahoa, Is. of Hawaii (toll-free 800-6-FLOWER or 800-635-6937) Also anthuriums.

For further information, consult "Flower Farms" in the Yellow Pages, or write to the Dept. of Agriculture, Market Development Branch, 1428 S. King St., Honolulu, Hawaii 96814.

CARE OF CUT FLOWERS

Heliconias

1. Wash in water with a little detergent and rinse with fresh water. Cut an inch off the bottom of the stem with a sharp knife, and soak entire stem and flowerhead in water at room temperature for ten minutes. If possible, resoak every three to five days; at least use a mister. These soakings simulate the frequent daily deluges that heliconia experience naturally.

2. An aerosol leaf shine adds to their luster and keeping qualities. Flowerheads will not open any further. *Floral preservatives or sugar solutions are ineffective,* as water uptake by heliconias is minimal; they only encourage the growth of stem-clogging bacteria.

3. Do not refrigerate. Below 55 degrees F heliconias suffer from cold injury.

4. Keep away from sunlit windows, heaters, air conditioners and drafty places. Maintain humidity as high as possible.

5. The smaller the heliconia, the shorter will be its vase life. The smallest last one week, the largest, three or more weeks.

6. Heliconia flowers, arising from within the colorful bracts, last only one day; if dead flowers accumulate, carefully cut off or rinse out.

Gingers, Anthuriums, and Assorted Exotics

1. Trim and soak as for heliconias at room temperature (red ginger one-half hour, shell ginger, anthuriums, and others ten minutes).

2. Remove the shell ginger's natural protective sheath.

3. Encourage humidity and discourage sunlight, air conditioning, or drafts. Otherwise, the bracts will dry and curl at the edges.

4. *Do not refrigerate.*

5. Soak bird-of-paradise 20 minutes. New flowers do not automatically emerge from their heavy protective sheaths, so insert both thumbs near the bract's narrow end and gently tease unopened flowers upwards. Tough as the flower is, its crisp tissues break easily, so be careful. Snip off the thin white membrane separating the flowers. Recut stems every two days. All three "birds" (including traveler's tree) respond well to preservative solutions available from florists.

6. Anthuriums: soak every three days and mist daily or provide a relative humidity of 60 percent. Floral preservatives may be used.

7. Store red gingers upright. A little sugar in the water helps.

Foliage

Trim off one-half inch of stem, soak as for heliconias and keep stems in at least two inches of water. Hold at 45 degrees to 55 degrees F if not used immediately. Mist daily, especially the thin-leaved species.

Drying Flowerheads

Tropicals such as rattle shakers, Indonesian wax, and calathea wheat dry well by hanging upside-down in a well-ventilated room. After drying, spray with a product such as Design Master Super Surface Sealer to enhance their appeal and durability.

A modern bouquet of heliconias and *ti* leaves. Floral art in Hawaii, in its many forms, is offered with the underlying oriental philosophy that it will enrich the lives of viewers.

FLOWER ARRANGEMENT

Arranging flowers is like playing a musical instrument . . . a mysterious inner energy warms us, flowing out through our minds and hands. It awakens our artistic nature and is fun, but it does require a few materials and a modicum of concentration. Most people are already attuned to proportion and color balance from activities such as clothing and make-up. These talents are easily applied to flowers. However, one's attitudes and immediate state of mind must be positive to create a harmonious work of floral art.

Many people do not know what to do with tropicals; one cannot merely stick them in a jar and expect them to look like an arrangement in a fashion magazine. Although some are large and heavy, it is easy to create a conversation piece with fresh tropicals, foliage, and the following items:

1. *Frogs.* These closely-packed metal spikes attached to a heavy solid base are indispensable. In contrast, heavy-duty floral foam breaks apart quickly. An initial investment in frogs will last a lifetime. Buy several sizes (up to five inches across) from a florist or garden supply store. Shop first by phone; they are not easy to find, especially in Hawaii. Homemade substitutes can be constructed from plywood and turned-up nails, using rocks as stabilizers. Floral wires may be used with no ill-effects as their water uptake is minimal. Short, sturdy sticks placed in the frog around a thick stem will hold it in place, especially if the stem is leaning. This technique is used extensively in Japan.

2. *Strong garden clippers or sharp knife.* Every few days recut the stems at an angle, preferably underwater. This allows water to pass upwards more freely. Trim off foliage that droops or interferes with the design. Do not squeeze the plant tissues excessively.

3. *Vases.* Big, small, tall, flat and wide, round, square . . . invest in whatever you like. Experiment with casserole dishes, mixing bowls, spray-painted cake pans and aluminum basins. Oriental import shops and contemporary pottery outlets usually have shapely containers. Vases must be scrupulously clean. Wash out scum from previous flowers and change water often. Tall vases are effective for pendent heliconias, while flatter, open vases are best for upright species. Vases with narrow tops are generally impractical, but a single fern and a couple of gingers will transform even the lowliest wine bottle.

4. *A large, plastic bucket* for soaking heliconias and foliage and mister for frequent spraying.

5. *Miscellaneous "cover-ups"* to hide the frog and lower stems: driftwood, smooth beach stones, crinkly lava rocks (though not from Hawaii's national parks, where Pele, the Polynesian goddess of fire, may inflict calamity on you!), eroded beach coral, etc. A small supply of dried grasses, twisted vine stems, figurines, attractive seed-pods, etc. are additional options for creating different textural contrasts, moods, and whimsical flourishes.

Styles and Ideas

Each combination of flowers and foliage presents a unique challenge to the arranger. No one way is right. The old-time Hawaiian style favors a mass of vivid color that dazzles the eye. The more you add the better. Strive for symmetry. Oriental traditions lean towards modernization of traditional *ikebana* (Japanese flower arrangement; *ikebana* means "living flowers") techniques: sleek, graphic, and simple. Here each element is trimmed and painstakingly placed to reveal its lines and color as cleanly as possible. Every shape or geometric pattern is savored individually. The three basic lines are cut proportionately: select an appropriate length for the tallest stem in relation to the container size and amount of space the final arrangement will occupy, then cut the middle stem two-thirds, and the smallest stem one-half of that length. Heliconias and gingers are ideal for these principal lines. Conventional flowers, smaller exotics, and foliage are added last. Favor asymmetry, simplicity and elegance. Do not clutter. Observe the natural growth patterns and try to create an art work even more beautiful than nature.

Also try the in-between route, prevalent in the gorgeous contemporary arrangements at island resorts. Trailing orchids, large proteas, and tropical foliage are topped by hanging heliconias or ornamental bananas. Such exotic elegance mingles Western, Eastern, and contemporary art and, in my estimation, is unsurpassed in any other tropical country.

Each flower and leaf exudes such inherent individual beauty that, however we arrange them, we may well take some tips from the Japanese, long-renowned for their sensitivity in floral art: handle each leaf gently, respecting its life; use foliage to soften the pattern of stems and flowers, embellishing their fluidity of line in such a way that you try to express love, beauty, and balance in their highest senses. Create a living picture — a three-dimensional painting in space — which, though fleeting, is capable of touching someone's heart.

Specific Examples

- Intermingle tropicals with temperate-zone flowers: The bold, upright exotic look of heliconias and gingers offsets the graceful curves and more delicate blossoms of daisies, chrysanthemums, dogwood, etc.
- Combine contrasts of texture and form: smooth with fuzzy, pointed with feathery, light with heavy, broad leaves with complex shapes, etc. For example: papyrus with upright heliconias, pampas grass with bird-of-paradise, velvety ornamental taro with golden shampoo ginger, baby's breath with calatheas or 'olena.
- Match textures and form: roadside grasses with "calathea wheat" and rattle shaker; zinnias with gingers.
- Combine tall elements with low, round ones: hanging heliconias with peonies, ornamental bananas with torch ginger, large heliconias with monstera leaves.
- Blend variegated foliage with solid colors: multicolored *ti* leaves with pink ginger, 'Tricolor' dracaena tufts with torch ginger and self-peeling banana.
- Mix multicolored heliconias with plain foliage: rainbow heliconia with 'Sharonii' foliage, red-and-yellow heliconia with day lilies and dieffenbachia foliage. For eye-catching simplicity, insert a small psittacorum and fern frond into a narrow-necked vase.
- Color coordinate for festive occasions: 'Jungle King,' red ginger or 'Holiday' heliconia, pine branches, and silver pinecones (Christmas); red anthuriums, lobster claws, and heart-shaped philodendron leaves (Valentine's Day); pink ginger or ice-blue calathea, baby's breath and papyrus (baby's birthday); 'Jacquinii' lobsterclaw, parrot psittacorum, and Song of India dracaena (Golden Anniversary); white calathea, 'Warneckii' dracaena, white carnations, and shell ginger (Silver Anniversary).

Obake anthuriums and red ginger are embellished with a touch of Christmas tree clubmoss *(Lycopodium cernuum)* and ginger leaves.

Red anthuriums, 'Holiday' heliconias, red ginger, *ti* leaves, and the silvery foliage of Indian summer banksia *(Banksia occidentalis).*

'Sharonii' heliconia, Indonesian wax ginger and philodendron leaves.

A box of tropicals, moments before shipping. Each flower originates from a different area of the world.

A beautiful display of orchids, anthuriums, pincushion and king proteas, pink ginger, vari-lauhala and *ti* adorns a resort foyer.

An elegant and contemporary arrangement utilizes yellow caribaea and 'Sexy Pink' heliconias, obake anthuriums, pink ginger, papyrus and cut heliconia leaves.

To be surrounded by floral color each day is a real privilege. A greater thrill is to know that these beauties will bring happiness to thousands of people. Read the section on caring for flowers to derive maximum enjoyment from your flowers.

An ikebana *objet d'art* utilizing 'Kamehameha' lobster claws, white mink proteas and assorted foliage, courtesy of Exotica Design Maui. Hawaii is an ideal place for practicing the art of flower-arranging.

Logging trucks in Costa Rica, today typifying developing countries, recall the pioneering days of America. Up to 1,000 species of living plants and animals may be killed with each tree that falls.

Preserved by planning, a wealth of heliconias, bromeliads, and orchids grow protected in a mossy elfin forest, which is continually shrouded in fog. Monteverde Cloud Forest, Costa Rica.

A CONSERVATION NOTE

This book discusses some of the most dazzling flowers and foliage in the world. Today, especially in Hawaii, we are privileged to witness their bounty in cultivation. The number of tropical plants in cultivation is, however, only a fraction of those remaining in the wild. Unfortunately, the luxuriant ecosystems that nurtured them for millennia are disappearing at an alarming rate, victims of overpopulation, poor economy and ignorance. Such large-scale destruction affects all of us: the air we breathe; the climate we experience; the water we drink; the medicines, food and clothing we may need; the vacation we may want; and our moral and religious convictions. Do we have moral or spiritual rights to destroy forests of indescribable beauty, with their tens of thousands of species of plants, birds, monkeys, frogs, lizards, butterflies, and other creatures?

Today more than half the world's equatorial forests have been logged or burned. Currently 20 million acres — the size of West Virginia — disappear each year. Many worldwide organizations are attempting to salvage the last remaining tracts of equatorial forests, before it is too late, by providing assistance to developing countries such as Belize, Costa Rica and Malaysia. All benefit from the imaginative schemes in which foreign debts are purchased, then traded back to them in return for their preservation of forest tracts. "Ecotourism" — tourism based wholly on the public enjoyment of tropical forests and their varied life-forms — is adding a new world of employment, education, and economic opportunities. Obviously this helps to preserve precious resources, including undiscovered tropicals, for ourselves and future generations. Land can be saved for as little as $20 per acre.

If you would like to help, U.S. tax-deductible contributions may be sent to **The Nature Conservancy**, 1785 Massachusetts Avenue NW, Washington, D.C. 20036 and **World Wildlife Fund**, 1250 24th St. NW, Washington, D.C. 20037. Earmark your gifts for their tropical ecosystem campaigns.

The *coqui,* an abundant tree frog endemic to Puerto Rico, rests on a heliconia within the Caribbean National Forest.

About the Author

Dr. Angela Kay Kepler, a naturalized New Zealander, was born in Australia in 1943. A writer, photographer, field biologist, biological illustrator, and environmental consultant, she holds degrees from the University of Canterbury (New Zealand), University of Hawaii (Honolulu) and Cornell University (New York). She also spent one year as a postdoctoral student at Oxford University, England. She lived and worked in a tropical rain forest in Puerto Rico for three years. She and her husband have also hiked extensively in Central and South America, West Indies, Southeast Asia and the Pacific, including Hawaii.

Kay first came to Hawaii as an East-West Center foreign student in 1964. Since then, she has authored a dozen books and numerous scientific publications, written newspaper columns on biological aspects of the Hawaiian Islands, and contributed regular articles and photos to island publications. She has also assisted her husband in forest bird and plant surveys, seabird studies, and endangered species research in the mainland U.S., Hawaii, West Indies, and New Zealand.

About the Photographer

Born and raised on Maui, Jacob R. Mau has recently become nationally and internationally recognized as an outstanding photographer by taking second place in a Sierra Club-sponsored photo competition which drew 19,000 entries .

His photographs, rich with color, life, and design, focus on Hawaii's exotic flowers. He never uses flash, studio, or other artificial lighting. Jacob and the author have been friends for ten years, have previously collaborated on a companion volume, *Proteas in Hawaii*, and miscellaneous magazine articles. Jacob's pictures also appear in the splendid, large-format coffee table book, Maui on My Mind (Mutual Publishing Co.), in a floral calendar (Hawaiian Resources, Inc.), in *Mauian* and *Maui, Inc.* magazines, and in resort publications. Notecards and posters of Mau's work produced by TROPICALS of Alii Gardens, Hana, Maui, are prized by art lovers and horticulturalists alike.

Brooke Bearg, a former resident of Maui now living in Washington, art directed many of the outstanding floral photos in these pages. She was also a consultant on the production of this book.

INDEX

MISCELLANEOUS EXOTICS

TROPICAL FOLIAGE

Other Titles by the Author

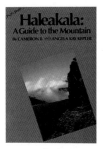

Haleakala: A Guide to the Mountain
by Cameron B. and Angela K. Kepler

The entire mountain, from sun-spangled shorelines through lush lowland forests, verdant pastures and alpine expanses. History, geography, cultural events and accommodations, a complete hiking and camping guide, points of interest, day trips and extended hikes. Over 200 color photographs, maps.
ISBN 0-935180-67-2 96 pages 5 3/4" X 8 1/2" $9.95

Proteas in Hawaii
by Angela K. Kepler
Photography by Jacob R. Mau

Floral photography at its best through the lens of award-winning, Maui-born photographer Jacob R. Mau. Over 200 photographs on this amazing flowering plant family.

Dr. Kepler's authoritative text provides a wealth of information on correct English and scientific names including pronunciation, buying and caring for the plants, flower arrangements, general and historic information. The first book on proteas ever published in the United States.
ISBN 0-935180-66-4 80 pages 5 3/4" X 8 1/2" $9.95

Maui's Hana Highway: A Visitor's Guide
by Angela K. Kepler

The incredible 52-mile journey of 617 curves and 56 bridges through some of Hawaii's most breathtaking scenery. Packed with hundreds of facts and interesting information.
ISBN 0-935180-67-2 80 pages 5 3/4" X 8 1/2" $9.95

HOW TO ORDER
Send check or money order with an additional $2.00 for the first book and .50 cents for each additional book to cover mailing and handling to: (Allow 3-4 weeks for delivery)

Mutual Publishing
1127 11th Avenue, Mezz. B
Honolulu, Hawaii 96716
Ph: (808) 732-1709 • Fax: (808) 734-4094
Email: mutual@lava.net